# A History of Dinosaurs in 50 Fossils

PAUL M. BARRETT

Smithsonian Books
Washington, DC

First published by the Natural History Museum, Cromwell Road, London SW7 5BD.
© The Trustees of the Natural History Museum, London 2024

The Author has asserted his right to be identified as the Author of this work under the Copyright, Designs and Patents Act 1988.

All rights reserved. No part of this publication may be transmitted in any form or by any means without prior permission from the publishers.

Published in North America, South America, Central America, and the Caribbean by Smithsonian Books

This book may be purchased for educational, business, or sales promotional use. For information, please write: Special Markets Department, Smithsonian Books, P.O. Box 37012, MRC 513, Washington, DC 20013

ISBN 978-1-58834-733-6

Library of Congress Cataloging-in-Publication Data available.

Printed in China, not at government expense
28 27 26 25 24             1 2 3 4 5


Internal design by Mercer Design, London
Reproduction by Saxon Digital Services, UK
Printed by Toppan Leefung Printing Ltd, China

# Contents

INTRODUCTION  4

## Setting the scene  8
A new group of reptiles 8 • What is a dinosaur? 12 • Origins 16 • The first dinosaur? 20 • Early dinosaurs 22 • The family tree grows 24 • Early hunters 28 • Lumbering long-necks 30 • Specialist herbivores 32 • Dinosaurs take over 34

## The key dinosaurs  38
Different tooth types 38 • Living tanks 40 • Plates and spines 44 • Head-bangers 46 • Thrills and frills 48 • Mesozoic cows 52 • Honking duckbills 56 • Dinosaurs go global 58 • The first giants 60 • Whip-tailed sauropods 64 • Big-nosed behemoths 66 • Crested hunters 70 • Southern predators 72 • Aquatic dinosaurs 74 • Jurassic hunters 78 • Feathered theropods 80 • The terrible claw 84 • Taking flight 88 • Early birds diversify 90

## Dinosaur biology  92
Scaly skins 92 • Feathers and fuzz 94 • Colour 98 • Soft tissues 100 • Dinosaur reproduction 102 • Parenting 106 • Sexing a dinosaur 110 • Life at extremes 114 • Growing up fast 116 • Ecto- or endothermic? 120 • Take a deep breath 122 • Speed and gait 124 • On all fours 128 • Predation 130 • Herbivory and gut bacteria 132 • Diet 134 • Brain size 136 • Hearing and sight 140 • Social behaviour 142 • A global phenomenon 146 • End of an era 148

GEOLOGICAL TIMESCALE  152
SPECIMEN DETAILS  153
INDEX  157
FURTHER READING  160
ACKNOWLEDGEMENTS  160
PICTURE CREDITS  160

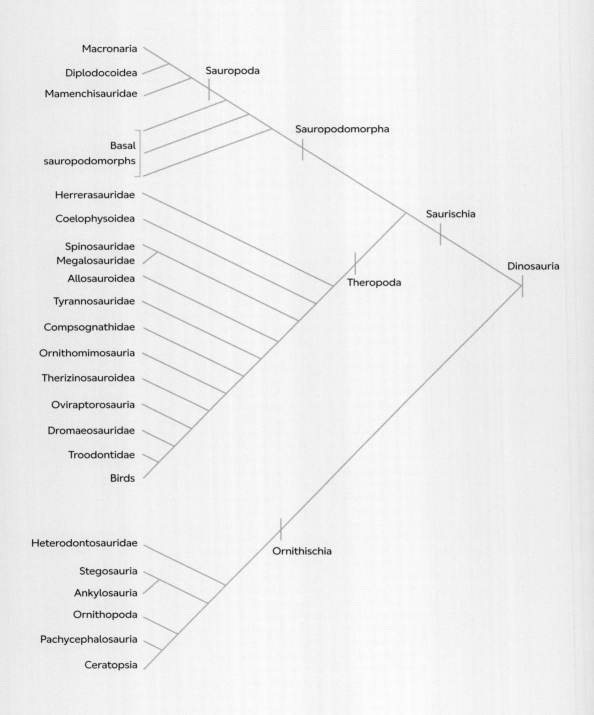

# Introduction

DINOSAURS HAVE FASCINATED AND EXCITED us for 200 years, ever since their fossilized remains first came to public attention. Images of these amazing animals now pervade popular culture, and people flock to museums to marvel at their skeletons. This interest is fuelled by the rapid pace of new discoveries, with announcements of new species or novel insights into dinosaur biology appearing on newsstands around the world almost daily. At least some of our interest in these ancient reptiles probably stems from the fact that dinosaurs look a lot like real-life versions of mythological monsters, such as dragons and basilisks. In common with those legendary creatures, dinosaurs reached enormous sizes, sported imposing horns and armour, and had jaws lined with dagger-like teeth. However, unlike dragons, dinosaurs are not figments of our collective imagination but were real animals that lived at a time when the Earth itself, as well as the animals and plants alongside them, looked quite different from now.

Scientists who study dinosaurs make new discoveries in the field, extracting previously unknown specimens from the ground. Other revelations come from surveys of historical museum collections or work in the laboratory. New technologies are continually applied to these old bones, including state-of-the-art imaging techniques, sophisticated chemical tests, and complex statistical analyses. In addition, the scientific study of dinosaurs began with only a handful of savants in nineteenth century Europe, but now there are hundreds of dinosaur specialists spread globally. As a result, we're living in a 'golden age' of dinosaur research.

As I write these words, we know that dinosaurs diversified into hundreds of species that lived all over the world. They occupied an impressive variety of ecological roles and pushed at the size limits that animals achieved on land. Dinosaurs even took to the air – overwhelming evidence shows that birds are the direct descendants of small, meat-eating dinosaurs – so there are 11,000 or so living species today.

Given their familiarity and popularity, dinosaurs provide a gateway to the other sciences and to a deeper knowledge of the natural world. Their antiquity gives insights into the depths of geological time and the processes that have transformed the Earth's surface and climate; understanding their behaviour and lifestyles requires us to apply knowledge gained from chemistry and physics, as well as biology; and the methods we use to study them draw on engineering, mathematics and computational modelling.

OPPOSITE: This evolutionary tree shows how the major dinosaur groups were related to each other. However, these relationships are much debated and some of the details are likely to change in future.

In writing this book, I've attempted to give an accessible introduction to the many facets of current dinosaur science, underpinned by the latest research in the field. Many of the examples I use simply weren't known to the authors of the books that I read as a dinosaur obsessed child in the 1970s. Indeed, dinosaurs have changed radically since even my earliest years as a professional palaeontologist – I began my PhD in 1993 before the first feathered dinosaurs had been discovered and before CT-scanning had become widely available.

The book is divided into three broad sections. The first section, pp. 8–37 explains what dinosaurs are and how they are related to other reptiles. Short biographies are provided for each of the major dinosaur subgroups in the second section (pp. 38–91), summarizing their defining features. Finally, various aspects of dinosaur biology, such as locomotion, reproduction, feeding and behaviour, are explored in more depth in the final section (pp. 92–151). Each entry can be read as a standalone essay, but they are all interlinked through the narrative of dinosaur evolution.

A single dinosaur fossil has been chosen as the focus of each chapter, serving as an exemplar to illustrate a particular aspect of dinosaur biology or classification. This isn't intended as a definitive list – other dinosaur specialists would have their own favourites – but it reflects my own interests and biases, which have evolved in turn as I've learnt more about these animals. Many are examples drawn from either my own research or the amazing collections of the Natural History Museum, London. The Museum's dinosaur collection is one of the world's most important, and includes some of the earliest dinosaur fossils to be found as well as examples of over 200 dinosaur species. It continues to grow as new specimens are added and I've had the privilege of working with this resource since I first encountered it as an intern in 1992. However, no museum collection is entirely comprehensive, so I've also used many examples housed in other museums around the world, reflecting the fact that dinosaur science is an international, team endeavour.

Dinosaurs dominated land ecosystems for nearly 150 million years. This scene shows one of the most famous examples, from the Upper Jurassic Morrison Formation of the USA, with *Stegosaurus* (left), *Diplodocus* (centre) and *Allosaurus* (right).

# A new group of reptiles
## *Megalosaurus*

THE VALES AND HILLS of the Cotswolds, in the heart of southern England, are dotted with postcard-pretty villages. Straw-thatched cottages built of warm, honey-coloured limestone form tight, cosy clusters around churchyards and inns. At first glance, the rocks used to build these sought-after properties do not look remarkable. However, a closer look reveals that they did not form from the geological processes that created the granites, sandstones and other building stones used elsewhere. They had a very different origin and owe their existence to biology – the tireless work of countless corals, clams and plankton. These unsung labourers, which thrived in their billions, were part of an ancient construction industry that prospered in the warm shallow seas covering western Europe during the Middle Jurassic period, around 174–163 million years ago. By extracting the calcium dissolved in seawater, these ancient engineers built strong, intricate skeletons. These structures accumulated, cemented and compacted into rock, trapping the shells and bones of the other creatures that lived and died alongside them. Layer after layer of this organic debris arrived on the Jurassic seafloor, each with its own distinctive appearance and composition, reflecting myriad differences in the habits of their animal builders, changes in the water chemistry and in the proportions of sand and mud added by the erosion of nearby islands. Millions of years later, quarrymen and builders found that some of these limestones were ideal for building cottage walls, whereas others split more easily into thin slabs that made perfect roofing tiles.

During the eighteenth and nineteenth centuries, another biologically derived rock – coal, the product of even more ancient swamp forests – fuelled the Industrial Revolution. This, in turn, led to further exploitation of the UK's natural resources, with economic growth stoking demand for goods, services and infrastructure. Mining for building stone, fuel and ore increased in scope and intensity. Although a farming area, the Cotswolds were also mined, with quarries opening to provide stone for the expansion of towns and cities, including the establishment of grand new colleges at the University of Oxford, the local seat of learning. Many of these building stones could be reached easily from the surface, but some – the layers providing the best roofing 'slates' – were in seams that were harder to reach. These required digging claustrophobically narrow tunnels deep underground, lit only by candlelight or oil lamp. This was dangerous work, as the tunnels often collapsed, but the miners excavated thousands of tonnes of slates, which still adorn the roofs of chapels and dining halls.

As the miners worked, they found strange objects similar to the bones of the horses, sheep and cattle living in the farms above, but differing in shape, and often of prodigious size. Some of these bones were acquired by local pastors and aristocrats who used their financial resources and ample leisure time to study these objects, thereby contributing to the developing sciences of biology and geology. This was a revolutionary time, when

This imposing lower jaw is one of the original finds used by William Buckland when he named *Megalosaurus* in 1824.

OPPOSITE: *Megalosaurus* was one of the earliest large predatory dinosaurs, reaching up to 9 m (29½ ft) in length. During the Middle Jurassic, central and southern England were flooded by a warm shallow sea and *Megalosaurus* would have hunted along these tropical shorelines.

the dogma of Christian scripture was being challenged by experiments in laboratories and observations in the field. These 'petrifactions', the objects we now call fossils, were already known to be the remains of once living creatures, but they were often unlike anything alive today. These curiosities fuelled vicious controversies on the age of the Earth, the processes that created our landscapes, and on the life of the past.

Dozens of large, fossilized bones emerged from the slate mines around the village of Stonesfield, Oxfordshire. Many of these were acquired by the Anatomy School, in Christchurch College at the University of Oxford. Several scholars worked on this material but it was the brilliant and eccentric cleric, the Reverend William Buckland, who first brought them to public prominence. Buckland, Oxford's first academic geologist, was initially stumped by the jaw bone, vertebrae and gigantic limb bones in these collections. However, the great French anatomist Georges Cuvier, who visited Buckland in 1818, applied his extensive knowledge of living animal anatomy to the problem, identifying the remains as those of a previously unknown gigantic reptile.

Cuvier's insights, and those of other colleagues, enabled Buckland's studies to progress and he presented the results of his work to a meeting of the Geological Society of London on 20 February 1824. During the proceedings he described in detail the remains of this '…enormous fossil animal…', which he identified as '…belonging to the order of Saurians or lizards'. Buckland painted an evocative picture envisioning this creature – dubbed *Megalosaurus*, the 'great lizard' – describing it thus: '…the beast in question [would have] equalled in height our largest elephants, and in length fallen but little short of the largest whales'. What Buckland did not know was that he had just named the first example of what would become one of the most famous animal groups of all time: he had named a dinosaur.

# What is a dinosaur?
## *Mantellisaurus*

ALTHOUGH MOST OF US would immediately identify *Tyrannosaurus* or *Diplodocus* as dinosaurs, what actually makes a dinosaur 'a dinosaur'? Films, television programmes, books and games often depict extinct animals like woolly mammoths, flying reptiles and marine reptiles alongside each other. As a result, they are sometimes accidentally labelled as dinosaurs too – but they very definitely are not.

Dinosaurs are reptiles, belonging to the same great cohort of backboned animals as crocodiles, lizards, snakes and turtles. Surprisingly, birds are also reptiles, as all of these different groups are descended from the same common ancestor, a small, lizard-like creature that scuttled through steamy swamp forests around 315 million years ago. All reptiles share a defining set of features inherited from this ancestor, as shown by both fossil and genetic evidence, but in some groups these characteristics were later overprinted or altered by evolution, which explains the many overt differences between feathery flying birds and their scaly-skinned crawling kin. An even more astonishing fact is that birds are not only reptiles but are embedded deep within the dinosaur family tree. Birds are the only members of this formerly great group to survive to the modern day.

Among reptiles, dinosaurs are characterized by a unique set of anatomical innovations that arose in the earliest members of the group, and which set them apart from their reptilian brethren. The most important of these are related to the way in which dinosaurs

This almost complete skeleton of *Mantellisaurus* exemplifies the novelty of the dinosaur body plan, with long hind legs that are tucked directly under the body. This innovation provided support and allowed for more efficient locomotion than in other reptile groups.

walk. In other reptile groups, such as lizards, the limbs jut out sideways and the body is suspended between them, in a low-slung posture. This means that the belly sometimes makes contact with the ground. In dinosaurs, something radically different occurred. The shapes of their thigh, shin and ankle bones changed, as well as the arrangement of their leg muscles. This enabled their legs to rotate inwards and downwards, bringing them directly beneath the body. This might sound unimportant, but it was one of the keys to dinosaur success. By being beneath the body, rather than at angles to it, the legs could act as pillars for support – a very efficient way of holding body weight against gravity because it requires little muscular effort to hold upright pillars in place, thus saving energy. In later dinosaurs, this enhanced body support would allow them to reach colossal sizes, an avenue that was closed to reptile groups that retained a splayed, sprawling stance. In addition, when lizards run or walk their legs have to swing through large arcs because they are pulled back and forth while the body wiggles from side-to-side. This is wasteful of energy, as each forward stride is much shorter than the full limb length. Dinosaurs, with their upright limbs, were able to walk much more efficiently, because their long legs could swing straight back and forth under the body, allowing much longer strides.

These changes were accompanied by the most startling dinosaur innovation – bipedality. Unlike other reptile groups, which remained on all fours (with some very rare exceptions), dinosaurs began their evolutionary journey by walking on their hind legs only. This was made possible by changes to their overall body proportions, such as a shorter trunk, and further modifications to their tails and hips, with the body balanced around the hips and hindlimbs. As all of the forces generated by walking now went through the hips, the linkage between the hips and the backbone was strengthened. In most reptiles, this relies on two specialized bones in the back, called the sacral vertebrae. These have expanded ribs that form solid connections with the hip bones. In dinosaurs, a third sacral vertebra was added, bolting the hips even more strongly to the rest of the body. Finally, the shift to bipedality also released the forelimbs for other uses, such as gathering food. Bipedality is an incredibly rare phenomenon in vertebrate evolution, seen in only a handful of groups, including humans. The vast majority of backboned animals living on land have remained firmly on all four legs. Other than humans, dinosaurs (including birds) are the only major group of animals to have been really successful in exploiting this peculiar way of moving around.

We can make some other deductions about the common ancestor of dinosaurs based on comparisons between the earliest known dinosaurs and their closest reptilian relatives. Like most other reptiles, this animal would have laid eggs rather than giving birth to live young. Also, the skulls and teeth of the earliest dinosaurs indicate that these animals were either meat-eaters or had mixed diets (including vegetation, insects and meat), so the first dinosaur was probably a carnivore or an omnivore. It is also possible that the first dinosaur was feathered, as many later meat-eaters were covered with dense plumage, but this is a controversial idea that is still the subject of much debate.

Taken together, this evidence suggests that the first dinosaurs were lightly built bipeds with slender necks and long tails; they walked on strong, upright hindlimbs and ate a variety of small prey, perhaps supplemented by plants. This ground plan was then shaped further by millions of years of evolution, eventually leading to a wide range of body shapes, with some dinosaurs adopting plant-based diets, others reinventing a four-legged stance and one group taking to the skies. It set the scene for the group to dominate life on land for over 150 million years.

*Mantellisaurus* comes from the Early Cretaceous period and its remains are known from southern England, Belgium and elsewhere. Its name honours Gideon Mantell, one of the pioneering scientists that helped to bring the first dinosaur fossils to light.

# Origins
## *Teleocrater*

DINOSAURS ARE MEMBERS OF a larger reptile group called Archosauria, the 'ruling reptiles', which is defined by the possession of several characteristic openings in the skull and jaw. One of these – the antorbital fenestra – is situated behind the nostrils and in front of the eye and is associated with a number of air-filled sacs. The other, a new opening in the lower jaw called the mandibular fenestra, forms an attachment site for powerful jaw-closing muscles. In addition to dinosaurs, Archosauria includes many other groups, including the flying reptiles and crocodiles. Fossil evidence suggests that the common ancestor of all archosaurs walked on all fours and had a number of crocodile-like features, such as a covering of bony armour plates. Unlike today's crocodiles, however, this ancestor lived on land. One of the major problems that palaeontologists have when studying dinosaur origins is in working out how an animal like this, which would have looked like a slender, short-snouted crocodile, could have transformed into a two-legged, long-necked runner.

Among archosaurs, the closest relatives of dinosaurs include the flying reptiles, or pterosaurs, and a group of small, agile runners called lagerpetids. Pterosaurs need little introduction – 'pterodactyls' are familiar characters in many films. They were the first vertebrates to evolve powered flight and soared through the skies as their dinosaur cousins foraged on the ground. A more obscure group, the lagerpetids were once thought to be dinosaur ancestors, as they share similarly long, slender hind legs. However, recent work has suggested that they were close relatives of pterosaurs instead. All three groups appeared in the fossil record during the Triassic Period (252–201 million years ago).

Pterosaurs and lagerpetids share many features with dinosaurs. Most of these relate to changes in hind leg anatomy, which is modified towards an upright stance in similar ways in all three groups. However, neither pterosaurs nor lagerpetids attain the full set of arm, leg, ankle and hip characteristics that define a dinosaur. Moreover, each of these groups has distinguishing features of its own, such as the development of the wing in pterosaurs. As a result, they do not provide much help in understanding how the quadrupedal ancestor of all archosaurs could have started changing into the agile biped that was the first dinosaur.

OPPOSITE: A hip bone (ilium) of the Middle Triassic reptile *Teleocrater*, which is thought to be an early dinosaur relative. Note the deep, bowl-like hip socket that gives the animal its name.

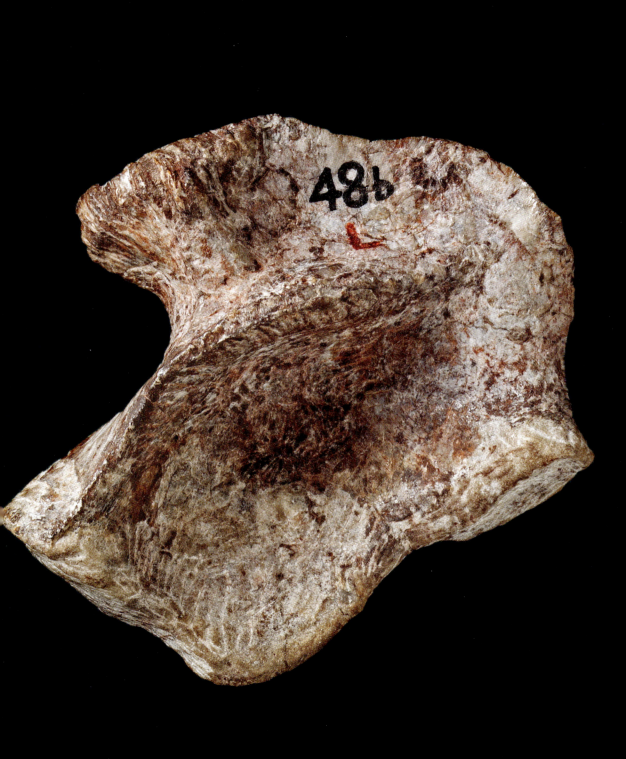

OPPOSITE: This scene shows *Teleocrater* (centre) with some of the other Middle Triassic animals it lived alongside. These include large, herbivorous dicynodonts (in the background), which are early members of the lineage that eventually led to mammals. Here, *Teleocrater* is feeding on a carcass of a near-mammal, a furry cynodont.

Luckily, a new chapter in this story opened recently with the naming of a new archosaur species from the Middle Triassic rocks (around 247–242 million years old) of Tanzania, east Africa. This animal, dubbed *Teleocrater rhadinus*, has helped to fill some of these evolutionary gaps. The first incomplete skeleton was found in 1933, but its importance went unrecognized for decades and it remained unnamed and unstudied. Much later, in 2015, the remains of at least three other individuals were discovered close to the original site. These new specimens prompted re-examination of the original skeleton. After detailed study it was realized that this animal narrowed some of the anatomical distance between the first dinosaurs and their archosaur ancestors. *Teleocrater* was finally named in 2017, more than 80 years after its remains were first uncovered. The name *Teleocrater* means 'complete bowl' in Greek and refers to the deep, bowl-like hip socket, an important feature in understanding hip and hind leg evolution, and *rhadinus* is the Greek word for 'slender' referring to the animal's gracile build.

When information from all of the available bones of *Teleocrater* was combined it became possible to reconstruct its overall appearance. Reaching up to 2 m (6½ ft) in length, including a long tail, *Teleocrater* had a skull with a low profile, a slender, elongate neck, a lengthy trunk, relatively short forelimbs and long, gracile hindlimbs. In many ways it looked like a mixture of a crocodile and a Komodo dragon. Crucially, however, it possessed a unique combination of anatomical features: some of these were identical to those otherwise found only in dinosaurs, whereas others were much more similar to those present in crocodiles and the archosaur common ancestor. For example, the ankle bones of *Teleocrater* have very complex shapes with complicated joints between them, similar to those of crocodiles and other archosaurs, but unlike the simple ankle bones of dinosaurs. Also, when the lengths of the forelimbs and hindlimbs are compared, their proportions are those of a four-legged quadruped, not a biped. However, other features link *Teleocrater* with dinosaurs. In contrast to earlier archosaurs, the first dinosaurs lacked bony armour plates in their skin: as far as we can tell, *Teleocrater* lacked these also. Dinosaurs also developed complex joints linking the individual bones in their spinal column, and *Teleocrater* possessed these features too.

When all of this anatomical information was included in an evolutionary analysis, *Teleocrater* was found to occupy a place in the archosaur family tree between the ancestral archosaur and dinosaurs. As a result, it helps to reveal some of the many

changes that had to take place in order to transform a crocodile-like ancestor into a dinosaur. For example, it shows that neck elongation and changes to the skull and backbone occurred before some of the key modifications to the hips and hindlimbs took place. In particular, the ankle and the hip socket of *Teleocrater* remain very crocodile-like. Only a single tooth has been found so far: it is pointed, curved and finely serrated, suggesting that *Teleocrater* was a predator.

The discovery of *Teleocrater* prompted reassessments of other similar, but less complete, fossils from around the world. Surprisingly, it was found that these more fragmentary remains, from India, Russia and Brazil, were from close relatives. This, in turn, raises the hope that other, more complete, fossils remain to be found in other Triassic rocks that will shed even more light on this major evolutionary transition.

# The first dinosaur?
## *Nyasasaurus*

THE DINOSAUR FAMILY TREE has deep roots, but determining how deep they go has been contentious. The geological ages of the rocks yielding the remains of *Teleocrater* and its relatives show that dinosaurs must have appeared sometime in the Middle Triassic (247–237 million years ago), or maybe slightly earlier. However, the earliest definite dinosaur fossils are from the Late Triassic (237–201 million years ago), more specifically from rock sequences in Argentina and southern Brazil that have been accurately dated to about 230 million years ago. However, one intriguing specimen, from the same Middle Triassic Tanzanian sediments as *Teleocrater*, offers a tantalizing glimpse of what might be the earliest dinosaur of all.

At first sight, the specimen is unimpressive: a single upper arm bone and a selection of broken vertebrae and ribs. However, on closer examination, these scant remains tell an important story. The upper arm bone, or humerus, bears a huge projection on its upper end that extends down the bone for more than one-third of its total length. This projection would have anchored large pectoral muscles, suggesting that the arms were strong and powerful. Three of the vertebrae show that they were involved in the complex attachment between the hips and the backbone, forming a structure called the sacrum. Finally, microscopic examination of a thin slice of bone, cut from the humerus, revealed bone textures consistent with extremely rapid growth. Among the many types of Triassic reptile that have been discovered so far, this combination of features is found only in dinosaurs.

This enigmatic specimen, named *Nyasasaurus parringtoni*, is thought to be 10–15 million years older than the other contenders for the crown of earliest-known dinosaur. Several evolutionary analyses have identified it as either a true dinosaur or the closest relative of dinosaurs yet found. However, this conclusion has been debated vigorously because of the fragmentary nature of the only known specimen and disagreements over the exact age of these Tanzanian Triassic rocks. Moreover, to date, most of the best early dinosaur specimens have come from South America, which has provided palaeontologists with many exceptionally preserved and well-dated skeletons, and it is generally assumed that this was the region in which dinosaurs took their first evolutionary steps. However, if correctly identified and dated, *Nyasasaurus* has the potential to push the search for the earliest dinosaurs further back in time. Together with *Teleocrater* and other Tanzanian discoveries, it also suggests that the African continent might have had a more important role in early dinosaur evolution than first thought.

Although fragmentary, this upper arm bone (left) and these sacral vertebrae (below) have features that are seen only in dinosaurs, suggesting that *Nyasasaurus* might have been the earliest-known member of this group.

# Early dinosaurs
## *Herrerasaurus*

THE LATE TRIASSIC WORLD hosted a bewildering variety of bizarre land animals, many of which have no living descendants. However, it was also the time at which many modern groups made their first appearance, including turtles, lizards, salamanders, frogs and mammals. If we took a Late Triassic safari, the commonest creatures we would encounter would include strange reptiles with no living analogues. There were ponderous herbivores, like the beaked, barrel-bodied rhynchosaurs and the spiky, armoured aetosaurs, as well as huge predatory crocodile relatives, called rauisuchians. Another group of vertebrates, commonly termed 'mammal-like reptiles' was also abundant. However, despite the name, these animals were not reptiles at all, but members of a distinct evolutionary branch, Synapsida, which split from reptiles more than 320 million years ago and which, ultimately, gave rise to mammals. Late Triassic synapsids included dicynodonts, sophisticated herbivores that range from rabbit-sized to hippo-sized, and a variety of small, perhaps furry, cynodonts, the closest relatives of mammals.

The first dinosaurs formed part of this ancient menagerie. However, their fossils are much scarcer than those of the other animals they lived with. By producing an accurate census of these fossils, palaeontologists have shown that dinosaurs began their evolutionary journey as rare animals that were bit-part players in these ecosystems, far from the starring roles they achieved later in their history.

In addition to their rarity, the earliest dinosaurs were small, with most reaching only 1–2 m (3¼–6½ ft) in length and with body weights of up to 30 kg (66 lb), the same as a large family dog. They were much smaller than many other Triassic reptiles and synapsids and were probably easy prey for their contemporaries. However, there was an exception to this rule, a giant by the standards of the time. This animal, *Herrerasaurus ischigualastensis*, from Argentina, reached up to 3 m (9¾ ft) in length and 350 kg (772 lb) in weight. *Herrerasaurus* is known from several skeletons, including one beautifully complete skull, so its anatomy is very well understood. As a result, it is often used as a 'road map' for understanding early dinosaur biology and for helping to identify and interpret less complete remains. Like all other early dinosaurs, it was a biped with long hind legs indicative of considerable agility. Its skull was robustly built, with a long snout, large eye sockets and rows of curved, razor-sharp teeth. The arms were powerful, each ending in a grasping hand with fingers tipped by large strongly curved claws. All of these features show that *Herrerasaurus* was an efficient, fearsome predator, perhaps the first specialist carnivore in dinosaur evolutionary history. The same features seen in *Herrerasaurus* were inherited by all later carnivorous dinosaurs, contributing to their success, which ultimately culminated in the evolution of the largest predators ever to live on land.

The long, low skull of *Herrerasaurus* was adapted for carnivory. A mobile hinge between the bones of the lower jaw helped it to deal with struggling prey.

# The family tree grows
## *Eoraptor*

ALTHOUGH THE FIRST DINOSAURS were rare, detailed examination of these remains shows that they diverged rapidly into different groups, each of which would go on to include the more familiar dinosaurs of the Jurassic and Cretaceous. These early splits in the dinosaur family tree, and ways in which these animals started to diverge from each other in appearance and behaviour, are captured in Late Triassic and Early Jurassic rocks around the world.

Within 5–10 million years of dinosaur origin, early representatives from at least two distinct dinosaur groups appear in the earliest Late Triassic rocks of Brazil and Argentina: Sauropodomorpha and Theropoda. The most iconic members of Sauropodomorpha are the gigantic sauropod dinosaurs, such as *Diplodocus*, which were to become important herbivores in later dinosaur ecosystems. However, the early members of this group – once termed 'prosauropods', and now formally 'basal sauropodomorphs' – were much smaller,

Many of the most complete early dinosaur skeletons, such as this *Eoraptor*, have been discovered in the arid badlands of Ischigualasto, in western Argentina. This is one of the most important sites there is for understanding the first stages of dinosaur evolution.

mainly bipedal and formed the ancestral stock from which the later giants evolved. The second dinosaur lineage, Theropoda, was specialized for carnivory and its most famous representatives, like *Tyrannosaurus* and *Velociraptor*, became the top predators of Mesozoic landscapes. In some ways, Theropoda was the most conservative dinosaur group, retaining many of the basic features seen in the dinosaur common ancestor, but during the Jurassic some theropods evolved feathers and gave rise to birds, exploiting a completely new way of life. Some dinosaur specialists think that there was a third, small group of early dinosaurs, called Herrerasauridae, including *Herrerasaurus* and a few Triassic relatives. If they represent a distinct group, then herrerasaurids might represent an early 'experiment' in specialist carnivory. However, most palaeontologists consider herrerasaurids to be early members of Theropoda, rather than belonging to a separate group of their own.

Another major dinosaur group appeared slightly later in the fossil record, with its first fossils coming from the dawn of the Jurassic Period, around 201 million years ago. This group, Ornithischia, eventually gave rise to animals like *Triceratops*, *Iguanodon* and *Stegosaurus*. The majority of ornithischians were herbivorous and many later types were quadrupedal. Nevertheless, the earliest members of the group were bipeds. Indeed, if you were transported to one of these early dinosaur ecosystems, it would have been hard to distinguish early theropods, sauropodomorphs and ornithischians from each other at a distance: most of these animals would have been small, bipedal, slender necked, long tailed, and either carnivorous or omnivorous. You would have to risk getting closer to look at details of the head, teeth, hands or feet to tell which kind of dinosaur you were looking at. This reflects the fact that these animals lived only a few million years after the dinosaur common ancestor, and had only just started to evolve the different, distinctive features that would eventually allow their descendants to be identified more clearly.

This leaves us with at least three major dinosaur groups (maybe four, if herrerasaurids are not theropods), all descended from the same common ancestor. But how are these groups related to each other? This question has vexed palaeontologists since the first dinosaurs were named, but the most popular solution was proposed by English palaeontologist, Harry Govier Seeley. In 1887, he noticed that theropod and sauropod dinosaurs shared many anatomical features, which led him to conclude that these two groups were more closely related to each other than either was to any of the

This beautifully preserved skull of *Eoraptor* contains a mixture of features associated with both plant- and meat-eating. As a result, it's generally considered to have been omnivorous. Other early dinosaurs probably had similarly mixed diets.

other dinosaurs. To capture this relationship, he coined the name 'Saurischia'. Saurischia means 'lizard-hipped', referring to the fact that theropod and sauropod hip bones are similar to those of living reptiles, with the front hip bone, called the pubis, pointing forwards. Seeley also noted other similarities between sauropod and theropod skeletons, such as the presence of distinctive air spaces in their vertebrae, which supported his idea. By contrast, Seeley placed the other dinosaurs then named, including *Iguanodon* and *Stegosaurus*, into a second group, Ornithischia, meaning 'bird-hipped'. Ornithischian dinosaurs are distinguished from saurischians by possessing a different hip structure: in ornithischians the pubis bone is rotated to point backwards (an evolutionary novelty that also occurred – independently – in birds, a source of much confusion that we will return to later, on p.86). Consequently, Seeley's classification established a major evolutionary split at the base of the dinosaur family tree – with a saurischian branch leading to theropods and sauropodomorphs, and a second branch leading to ornithischians.

Almost all palaeontologists have supported Seeley's ideas, and his scheme has been used to classify dinosaur species for well over a century. However, some specialists have proposed alternative classifications. One of these unites all plant-eating dinosaurs (Sauropodomorpha + Ornithischia) into a group called Phytodinosauria (meaning 'plant dinosaurs') to the exclusion of theropods. Another alternative has linked ornithischians and theropods into a group called Ornithoscelida (meaning 'bird-like arms') that excludes sauropodomorphs. Both alternatives remain controversial, however, and Seeley's original scheme remains the most popular, at least for now.

Regardless of the classification used, many areas of uncertainty and debate remain. In evolutionary analyses of early dinosaurs, some species – like *Eoraptor* from the Late Triassic of Argentina – flip back and forth between different groups, as they have a mixture of features that confounds their relationships. In addition, all of these classifications predict that ornithischians should have appeared at the same time as the other dinosaur groups – but their fossils are currently unknown from Triassic rocks. We do not know why.

# Early hunters
## *Coelophysis*

THE FIRST THEROPOD DINOSAURS established a successful body plan that endured for the entire Mesozoic Era. All theropods maintained the bipedal stance of the earliest dinosaurs and, with only a few exceptions, the majority of non-bird theropods also retained their carnivorous habits. They possessed a suite of skeletal features that distinguish them from the other major dinosaur groups, with many of these changes related to feeding and locomotion. For example, theropods had a hinge-like joint between the bones of the lower jaw that allowed the skull to absorb more effectively the forces generated by struggling prey. In addition, the number of toes on their feet was reduced from five to three, giving them a typically bird-like foot, with each toe tipped by a large curved claw. Theropods also possessed a wishbone, a V-shaped bone in the chest that was formed by the fusion of two shoulder bones, the clavicles. This structure acted as a flexible spring for the attachment of various chest and arm muscles, and this might have been important for forelimb function, as the arms were used for grappling with prey.

Early theropods are exemplified by species like *Coelophysis bauri*, known from the Late Triassic of the western USA. Unusually, *Coelophysis* is known from hundreds of well-preserved skeletons, most of which are from a single quarry in New Mexico, situated on the rather spookily named Ghost Ranch. It is not known why so many *Coelophysis* were preserved in the same place, but it has been suggested that they might have congregated at a water or food source and were overcome by a flash flood. Thanks to the abundance and quality of these skeletons, it is one of the best-known early meat-eaters, and some even have fossilized gut contents that reveal their last meals (including a small crocodile in one case). *Coelophysis* had many features that are common to all early theropods, including a lightly built, narrow skull with a long, pointed snout and large eye sockets (presumably indicating good eyesight, which would have been useful in locating prey). Its jaws were lined with rows of up to 27 backwardly curved, finely serrated teeth ideally suited for slicing flesh. The neck was slender and flexible, and the arms, ending in four-fingered, claw-tipped hands, were long and would have been used for manipulating food. Additional prong-like joints between the bones of the back helped to keep the body rigid during running, and the base of the tail acted as an anchor for efficient leg muscles that powered its long, elegant hind limbs, making it an efficient, speedy runner. Like other early theropods, *Coelophysis* was small, up to 3 m (9¾ ft) in length, and lightly built, weighing no more than around 25 kg (55 lb), the same as a labrador.

The name *Coelophysis* means 'hollow form' a reference to its lightly built skeleton and hollow, bird-like limb bones.

# Lumbering long-necks
## *Plateosaurus*

SAUROPODOMORPHS WERE THE first dinosaurs to achieve a truly global distribution and to appear in large numbers. Although rare at the onset of the Late Triassic, they increased in abundance and diversity throughout this period until, around 210 million years ago, they became the commonest large land animals in many parts of the world. In addition to their rapid geographical spread, they developed a variety of ways in which they fed and walked around. The very earliest members of the group, like *Buriolestes* from Brazil, from around 230 million years ago, were small, agile, bipedal and carnivorous, like other early dinosaurs. However, sauropodomorphs evolved rapidly, diversifying into a wide variety of different species, some with much larger body sizes (including the first dinosaurs to weigh 1 tonne (1.1 tons) or more), and others with different diets and new means of getting around.

Many of the features defining the group can be seen in *Plateosaurus*, which is known from dozens of skeletons excavated from rich quarries in the Late Triassic deposits of Germany and Switzerland. Perhaps the most conspicuous of these features relates to head size: sauropodomorph heads were very small with respect to overall body size. In fact, all sauropodomorph skulls measure less than half of the length of their thigh bones, and in many cases are even smaller, so their heads can appear absurdly tiny (think of how small the head is in *Diplodocus*, a later member of the group). Through evolutionary time, sauropodomorph necks increased in length, a feat accomplished through the addition of extra neck vertebrae and the elongation of each individual vertebra, thereby increasing their overall reach and range. Their large, five-fingered hands had an enormous, hooked claw on each thumb, which was probably used for defence, fighting or collecting food (and possibly all three). The skulls were lined with rows of many tall, thin teeth, which were ideal for biting through plants (and maybe the occasional small animal) and they had chunky, barrel-shaped bodies, long hindlimbs and powerful muscular tails. The majority of Triassic and Early Jurassic sauropodomorphs were bipedal, as shown by their limbs and trackways. They also started to exhibit new and interesting behaviours – animals like *Plateosaurus* were some of the first dinosaurs to be sociable, living in herds.

Just before the end of the Triassic Period, a few early sauropodomorphs started to develop larger body sizes and began to walk on all fours. These animals were the ancestors of the sauropods. Sauropods took these trends to extremes, eventually becoming the largest land animals of all time. The bipedal sauropodomorphs died out during the early stages of the Jurassic Period, but the sauropods went on to diversify in the wake of their demise, giving rise to hundreds of different species that reached almost every major landmass and thrived until the end of the Cretaceous.

*Plateosaurus* is one of the largest Triassic dinosaurs, reaching lengths of up to 10 m (33 ft). It is just one of many early sauropodomorph species and these animals lived all over the world, ranging from Greenland to Antarctica and from Argentina to China.

# Specialist herbivores
## *Lesothosaurus*

IN MANY WAYS, ORNITHISCHIANS are the oddest dinosaurs. Although ornithischians, theropods and sauropodomorphs inherited the same set of basic features from the dinosaur ancestor, in ornithischians this formula was taken and stretched in startlingly different directions. They are the most diverse non-bird dinosaurs in terms of appearance, with the group including animals as varied as the horned *Styracosaurus*, the armoured *Ankylosaurus* and the trumpeting *Parasaurolophus*.

*Lesothosaurus*, from the Early Jurassic of southern Africa, shows how even the earliest ornithischians had radically altered bodies in comparison to those of the other dinosaur groups. Its small, boxy skull had many new features, most of which were related to eating plants. These include the development of a bizarre, arrowhead-shaped bone at the tip of the lower jaw, called a predentary. During life, this bone – unique to ornithischians – was covered by a tough sheath of keratin (the same substance that our fingernails are made from), which produced a sharp-edged beak that was ideal for nipping at shoots and leaves. The teeth of *Lesothosaurus* were low, triangular and roughly serrated, all features that made them useful for slicing through vegetation (and sometimes small animals). A rod-like bone, called the palpebral, crossed the eye socket and might have supported the eyelid – palpebrals are present in all ornithischians, but were absent in other dinosaurs.

Small, bipedal and agile, *Lesothosaurus* (at up to 2 m (6½ ft) in length) was a speedy runner. Its hips, and those of all other ornithischians, were profoundly different from those of the other dinosaur groups. Not only had the pubis bone rotated from its usual forward-facing position to point backwards, but the front end of the upper hip bone (the ilium) had gained a long, strap-like prong. These modifications to the hips suggest that ornithischians were altering the arrangement and functions of their leg and trunk muscles, although we are not sure why. It is possible that the backward rotation of the pubis was related to their herbivorous diets, as this change created more space in the body cavity that could have been packed with a longer gut. Longer guts would have helped these small animals to digest their food for longer periods and more efficiently.

Later ornithischians kept and modified these attributes, especially as they developed ever more sophisticated ways to chew plants (no ornithischians were dedicated meat-eaters) and experimented with ways in which they moved around. For example, although some ornithischians remained bipedal, many evolved to become quadrupeds with further changes occurring in their hip, hand and leg anatomy. Ornithischians were among the most sociable dinosaurs, with numerous species known to live in herds (including *Lesothosaurus*), and some of their oddest features – such as the bony horns, frills and crests on their skulls – probably evolved for communication and display. Although they were not particularly common at the beginning of the Jurassic, ornithischians eventually became the most abundant dinosaurs of many Cretaceous ecosystems.

This three-dimensionally preserved *Lesothosaurus* skull is only around 12 cm (5 in) in length and is probably from a juvenile individual. It was recently CT-scanned to reveal details of its inner structure.

# Dinosaurs take over
## *Massospondylus*

THE END OF THE TRIASSIC WAS, quite literally, a disaster. It was marked by a series of dramatic changes to life on land and in the sea, and many groups of animals and plants that had been widespread and abundant for millions of years perished, never to be seen again. This end-Triassic event was so severe that it is classified as one of the 'big five' mass extinctions that punctuate the history of life on Earth, and over 70% of all animal species disappeared at this time. Although scientists have proposed various causes for this catastrophe, there is now broad agreement that it was the result of a global geological event that had devastating consequences. During the Triassic, all of the Earth's continents were joined together in a single land mass, Pangaea. However, the end of the Triassic witnessed the beginning of the end for this vast supercontinent. A huge plume of superheated rock started to rise from the Earth's interior underneath the northern part of Pangaea. This hot rock eventually found its way to the surface, and as it cooled and turned to lava it solidified at the surface and began to push what are now Europe, North America and Africa apart from each other. This tore northern Pangaea in two, creating a new ocean in the process – the North Atlantic Ocean – which has continued to widen ever since. The birth of the North Atlantic had three major effects on life. First, the gases given off by these volcanic eruptions affected the chemistry of the atmosphere and oceans, leading to intensive global warming and making seawater more acidic. The increased temperatures made life on land difficult and the acidic seawater was bad news for shell-building animals, like clams, whose shells do not form properly (or at all) in such conditions. Second, the new ocean introduced a lengthy strip of water into what was formerly the arid interior of Pangaea. This had another major effect on climate, changing rainfall patterns all over the globe. Finally, it physically separated North America from Africa and Europe, so that animals that had previously been able to pass back-and-forth between these areas were now separated. This three-way punch led to the extinction of entire groups and the dramatic thinning of many more. Life in the Jurassic Period would be very different.

However, it was not all doom and gloom. Some groups not only clung on to survive the extinction, but went on to prosper. Extinctions are not only endings but opportunities. Dinosaurs were one of the main beneficiaries. Although many individual dinosaur species died out during the extinction, all of the major dinosaur groups had survivors that passed through into the Jurassic. These pioneers went on to found the great dinosaur dynasties that dominated the landscape for the rest of the Mesozoic Era. Palaeontologists still argue about why dinosaurs survived when so many other reptiles died out, because it is not clear what features of dinosaurs enabled them to pass through this cataclysm. Was their fast growth an advantage? Did their efficient, speedy locomotion help? Was their dietary flexibility the key? Was it a combination of some, or all, of these factors? Or something we have not yet thought of? Frankly, we do not know, and palaeontologists are still working towards an answer.

Whatever the reason, dinosaurs flourished in the wake of the end-Triassic extinction. The first years of the Jurassic witnessed the appearance of many new dinosaur species and groups, and dinosaurs became truly global for the first time, being present on every major land mass. They increased in numbers everywhere and quickly became the most visible and abundant land vertebrates in all Jurassic land ecosystems. This change is exemplified in one of the areas where I carry out fieldwork, southern Africa. During the Late Triassic, dinosaurs were present and becoming important, but they still shared the landscape with many other types of large animals, most notably large synapsids and some predatory crocodile relatives called rauisuchians. However, in the Jurassic rocks that lie on top of these deposits, which were laid down after the extinction, the only large animals present are dinosaurs. These earliest Jurassic dinosaurs include the basal sauropodomorph *Massospondylus carinatus*, a medium- to large-sized omnivore (up to 5 m (16½ ft)) known from the remains of hundreds of individuals. While herds of *Massospondylus* roamed the landscape, ornithischians also began to emerge as an important group, living alongside their sauropodomorph and theropod cousins.

From the beginning of the Jurassic Period 201 million years ago – indeed, until the end of the Cretaceous – dinosaurs occupied all the top herbivore and predator niches on land and diversified into many different shapes, lifestyles and sizes. For example, for the remainder of the Mesozoic Era, which lasted for another 135 million years, all land animals larger than a badger were dinosaurs. It seems that, rather than suffering badly from the effects of the extinction, dinosaurs took advantage of the situation, expanding to fill the many ecological roles left vacant by the disappearance of their predecessors. This has led some palaeontologists to suggest that dinosaurs were simply 'victors by accident', taking over through sheer luck rather than due to any biological features that might have promoted their success. Whether luck was the most important factor, or whether dinosaur biology was the key to their later dominance, is still heavily debated.

*Massospondylus* was the most abundant dinosaur of the Early Jurassic period. Palaeontologists have unearthed *Massospondylus* eggs, nests and embryos, as well as skeletons of hatchling through to adult size. As a result, this species is one of the best studied early dinosaurs.

# Different tooth types
## *Heterodontosaurus*

FOLLOWING THE END-TRIASSIC EXTINCTION, many different dinosaur groups began to diversify. Among these was a group of ornithischian dinosaurs that retained many features of the dinosaur common ancestor, but that had started to develop new, more sophisticated, ways of eating plants. These animals, called heterodontosaurids, were most abundant in the Early Jurassic of southern Africa and their best-known representative is *Heterodontosaurus*.

As with many other early dinosaurs, *Heterodontosaurus* was small, reaching no more than 2 m (6½ ft) in length, and was a slender-legged, long-tailed, fleet-footed biped. It lacked armour and its main defence would have been a speedy getaway. Its long, grasping hands were surprisingly similar to those of predatory dinosaurs, but its most distinctive features were in the skull. *Heterodontosaurus* means 'different-toothed reptile' and refers to its sets of chisel-like, closely packed, high-crowned teeth that occupy most of the jaws, in combination with a pair of dagger-like canine teeth near the front of the mouth. The chisel-like teeth ground against each other during chewing, wearing their tips flat, a pattern indicative of plant-eating. Large spaces at the back of the skull were occupied by bulging jaw muscles, showing that powerful jaw closing motions were possible. By contrast, the curved canines, which had fine, knife-like serrations, might have been useful for impaling prey. This combination of different tooth types, together with its grasping hands, might indicate a mixed, omnivorous diet in *Heterodontosaurus* – perhaps it changed diet depending on the seasons or it needed both plant and animal food to survive. When *Heterodontosaurus* was alive, the climate in southern Africa was arid, and water (and therefore plant) availability would have varied greatly throughout the year. Maybe a mixed menu helped *Heterodontosaurus* to prosper in this tough environment.

It is also possible that the canines served other functions, including display. Today, some forest deer, like muntjac, have sharp canine teeth that they use when fighting over mates and territory. Perhaps *Heterodontosaurus* used its canines for the same reason. Some scientists have suggested that only males had these teeth, but we do not yet have enough fossils to test this idea. *Heterodontosaurus* also had small triangular 'horns' jutting out from the sides of its skull: these might also have been for display.

Other heterodontosaurids reached the UK, USA, China and Argentina, and the last-known members of the group are from the Early Cretaceous. However, although they spread widely, and lived for many millions of years, they were uncommon animals and only a handful of species have been named. Heterodontosaurids are currently thought to be good models for understanding the anatomy and biology of the earliest ornithischians, as they seem to be very close to the base of the ornithischian family tree. However, their precise relationships with later ornithischians are still debated.

*Heterodontosaurus* had a robust, box-like skull that was adapted to withstand high bite forces. The long, fang-like canine on the lower jaw fits neatly into a notch on the upper jaw when the mouth is closed.

# Living tanks
## *Scolosaurus*

SOME ORNITHISCHIANS EVOLVED heavy-duty protection that would have been a major deterrent to all but the hungriest carnivores: namely, an extensive covering of impenetrable bony armour. All armoured ornithischians belong to a group named Thyreophora – a name meaning 'shield bearers' in reference to their protective coats. The most primitive thyreophoran, *Scutellosaurus*, was a small, herbivorous biped from the Early Jurassic of the USA. It looked a lot like other early ornithischians in most respects, but with one major exception: its back was covered with hundreds of small bony plates, termed osteoderms, which would have been arranged in parallel rows stretching over the animal's back and sides. The presence of these osteoderm rows is the main defining feature of all thyreophorans. Osteoderms form within the skin and are not attached to the rest of the skeleton. The outer surface of the osteoderm is usually covered with a large horny scale. Similar armour has evolved independently in many other reptile groups, such as crocodiles, and it gives them their characteristic gnarly appearance.

This *Scolosaurus* skeleton is notable for the completeness of its armour, which extends from the neck (towards the right) all the way over the back to the tail, forming an unbroken, impenetrable covering. It is among the most complete ankylosaurid specimens known but sadly lacks the skull.

The first definite thyreophorans appear during the Early Jurassic, with fossils known from the UK, Germany, USA and China. A few species from the Early Jurassic of South America and Africa, including *Lesothosaurus*, might also be members of this group (as they share a few features with thyreophorans), but these southern animals lacked armour (at least none has yet been found) so their inclusion is controversial. During the Early Jurassic, thyreophorans evolved rapidly and there were many changes to the ways in which they walked and fed. With the exception of *Scutellosaurus*, all other thyreophorans became quadrupedal, giving up a life of speed for majestically marching around on all fours. *Scelidosaurus*, from the UK, was one of the first thyreophorans (indeed, one of the first ornithischians) to become quadrupedal, as shown by changes in the anatomy of its hands and forelimbs, which morphed into pillars for support. As they went down on to all fours, thyreophorans also became larger – dainty *Scutellosaurus* tipped the scales at around 5 kg (11 lb), whereas an adult *Scelidosaurus* weighed closer to 250 kg (550 lb). In addition, thyreophorans abandoned all remnants of their omnivorous ancestry and became specialist herbivores, with numerous changes to the teeth, skull and jaw muscles that assisted in processing tough plant food, like cycads and horsetails. These include teeth that could grind against each other and larger jaw muscles that were capable of generating more powerful bites.

Sometime in the Middle Jurassic, Thyreophora split into two branches, each of which would go on to be important at different times in later Mesozoic ecosystems. These two groups include the most familiar armoured dinosaurs – the ankylosaurs and the stegosaurs (see p. 44) – and, although they were close cousins, each took quite different evolutionary journeys.

Ankylosaurs, like *Scolosaurus* from the Late Cretaceous of Canada, were the living tanks of the Jurassic and Cretaceous. They took the basic thyreophoran body plan, with numerous armour rows, and elaborated it, becoming some of the most extensively armoured land animals of all time. Many more rows of armour were added to their backs and flanks, and these were often interlinked to form a continuous bony covering. More rows of armour extended along the tail and on to the limbs. Armour plates were fused to the underlying bones of the skull and jaws. At least one ankylosaur had armour-plated eyelids, and some, such as *Ankylosaurus* from the latest Cretaceous of North America, developed massive tail clubs, which formed from huge osteoderms that fused with each other and with the vertebrae at the end of the tail. Powerful muscles could swing the tail from side-to-side, making these clubs a formidable defensive weapon when aimed at the shins of attacking theropods.

A close-up on the armour of *Scolosaurus*, showing some of the variation in individual osteoderm size and shape. On this part of the body the osteoderms are tightly packed, forming a continuous mosaic-like covering.

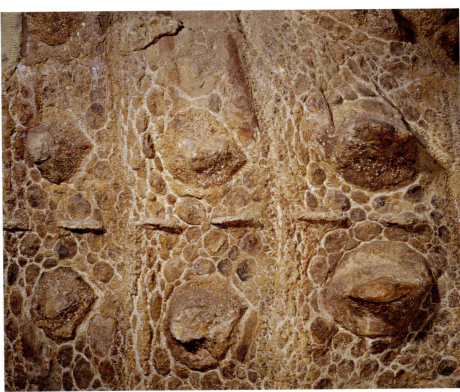

*Scolosaurus*, as viewed from below (with the neck end to the right). Note the wide hip region and the remains of one hindlimb tucked up beneath the body.

The only unarmoured part of an ankylosaur was its underbelly, but as this was held close to ground (and many ankylosaurs were pretty heavy, so hard to flip over), this vulnerable spot would have been hard for predators to reach.

All ankylosaurs were quadrupeds and none were speedy runners, relying instead on armour for defence. All were herbivores, having jaws equipped with small, leaf-like teeth that would not have been great for chewing, but would have been useful for slicing and nipping leaves. Most digestion was left to their long guts, housed in very wide, barrel-shaped bodies. If you had looked down on top of an ankylosaur, you would have seen that they had wide hips and long, strongly-bowed ribs that framed their rather squat, chunky torsos. Ankylosaurs occupied a range of body sizes, with the largest – *Ankylosaurus* – reaching over 8 m (26¼ ft) in length and maybe 8 tonnes (8¾ tons) in weight. Despite this, all ankylosaurs fed close to the ground, as their low-slung bodies and short necks did not allow them to reach much higher. We have many examples of ankylosaur footprints in the fossil record, and even have a few specimens, like the spectacular *Borealopelta* from the Early Cretaceous of Canada, which have remnants of their last meals preserved inside. We have not yet found any ankylosaur eggs or nests, so the way they grew up remains unknown, but some fossils hint that the youngest ankylosaurs might have lacked fully formed armour, which might have developed and become more extensive as the animals matured.

Ankylosaurs were rare during the Jurassic, but became abundant and widespread during the Cretaceous, with many different species appearing, some of which were thriving until the end-Cretaceous extinction. Their fossils are found all over the world from Antarctica to Canada and from Chile to Japan. This contrasts with their cousins, the stegosaurs, whose story was quite different.

# Plates and spines
## *Stegosaurus*

THE WORLD'S MOST COMPLETE *Stegosaurus*, nicknamed 'Sophie', greets people as they enter the Natural History Museum in London. *Stegosaurus*, from the Late Jurassic of the USA, is the best-known member of the Stegosauria, an armoured dinosaur group that includes only a dozen or so species. Although close relatives of ankylosaurs, stegosaurs differ from them in several ways, including a different body shape, with narrow, long-snouted skulls mounted on a slender neck and a very deep, narrow body. Stegosaur hind legs were much longer than their forelimbs too, so their backs were very strongly sloped, giving them a rather awkward, ungainly appearance. However, the most obvious difference is in their armour. Other thyreophorans had multiple osteoderm rows embedded in their skin, but stegosaur armour was drastically reduced, so that only a single pair of rows, placed either side of the backbone, remained. In addition, the shapes of these osteoderms contrasted sharply with those in other thyreophorans. Ankylosaur armour consisted largely of thick, flat plates and short, stubby triangular spines. By contrast, stegosaur osteoderms were larger, flatter, vertically held plates or long conical spikes. Differences in the shapes, sizes and positions of these osteoderms are one of the main ways in which stegosaur species are told apart from each other. In *Stegosaurus*, some of the plates were very tall (up to 1 m (3¼ ft)) and diamond-shaped in outline, and there were two pairs of large spikes at the end of the tail. Other stegosaurs, such as *Kentrosaurus* from the Late Jurassic of Tanzania, had different types of plates and spikes – in this case the plates were much smaller and in some parts of the body there were spikes instead of plates. A few species, such as *Gigantspinosaurus*, from the Late Jurassic of China, had massive, curved spines that jutted sideways from their shoulders.

The functions of stegosaur armour are much debated. It would have been useful for showing off to other stegosaurs, maybe in displays over mates or territory, and the many sharp spikes probably played important defensive roles. It is easy to imagine how a swipe from a spiked stegosaur tail would put off all but the most determined theropod. The large plates of *Stegosaurus* were probably too thin to be useful in defence, but might have played a part in regulating body temperature, by acting as living radiators that absorbed and dumped heat.

It used to be thought that the plates in *Stegosaurus* were paired, but complete specimens like 'Sophie' show that they alternated from side-to-side along the back. Plates on the neck and at the end of the tail were relatively small while the largest plates were situated over the hips.

Although instantly recognizable, stegosaurs were not as widespread or as common as ankylosaurs. They were important in the Late Jurassic, when they are known from the USA, Europe, China and Africa, but declined quickly in the Cretaceous, dying out long before the end of the Mesozoic. It is possible that they were driven to extinction by competition with their ankylosaur cousins, which increased in importance as stegosaurs declined, or it could be that a key stegosaur food source, such as cycads, became harder to find.

# Head-bangers
*Pachycephalosaurus*

THE DOME-HEADS, OR PACHYCEPHALOSAURS, are some of the most mysterious dinosaurs. They appear suddenly in the fossil record, near the end of the Cretaceous Period, and are known only from Asia and North America. Only around 15 species have been named and little is known of their early evolution. Most pachycephalosaur remains consist of broken partial skulls and much of our knowledge of their anatomy comes from a handful of more completely preserved specimens.

The largest, most famous pachycephalosaur is *Pachycephalosaurus* itself, from Canada and the USA. Reaching up to 5 m (16½ ft) in length, its most impressive feature was its skull, which bore a massive, solid bony dome above the eyes that reached up to 25 cm (10 inches) in thickness. The edges of the dome were studded with an array of small bumps and spines, giving it a knobbly, spiky appearance. Pachycephalosaur teeth were low and triangular, showing that they were herbivores, and the few complete specimens we have confirm that they had long hind limbs and much shorter forelimbs, showing that they were bipedal.

All pachycephalosaurs had thickened skulls, with a bony shelf sticking out from the rear margin. This shelf usually bore a variety of knobs or short horn-like spikes, similar to those in *Pachycephalosaurus*. However, only a few pachycephalosaur species had the thick, rounded, bulbous domes for which the group is most famous. Differences in the shapes, numbers, position and sizes of these bony ornaments are used to distinguish the various species from each other. There has been continual debate about the use of the domes and spikes, with earlier generations of scientists suggesting that they might have used the domes for head-butting like mountain goats – either for competing with each other, in disputes over mates perhaps, or for warding off predators. However, detailed analyses of the dome's bone structure using cutting-edge imaging techniques suggest that direct head-on-head ramming behaviour was not likely: the domes do not seem to be strong enough to withstand major impacts, and their rounded shape would mean that the clashing heads

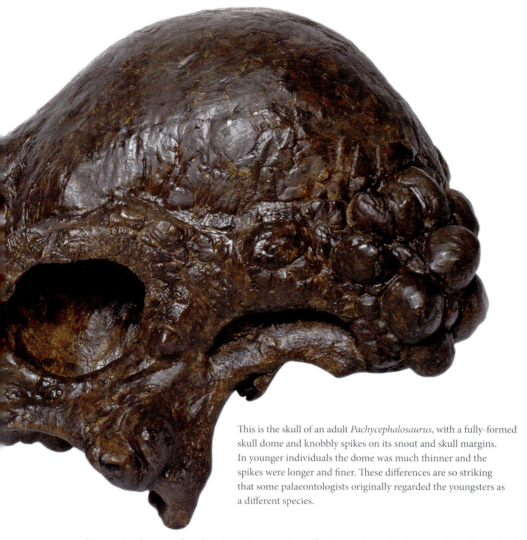

This is the skull of an adult *Pachycephalosaurus*, with a fully-formed skull dome and knobbly spikes on its snout and skull margins. In younger individuals the dome was much thinner and the spikes were longer and finer. These differences are so striking that some palaeontologists originally regarded the youngsters as a different species.

would simply glance off each other. Despite these findings, it is still thought that these skull features were important in behaviour, with the spikes being useful for visual display and the domes and flat shelf-like areas used for slower shoving contests, or perhaps for butting into the sides of opponents.

One other ornithischian group, the ceratopsians (horned dinosaurs), also developed bony shelves at the back of their skulls. The shared possession of this feature is often used to link pachycephalosaurs and ceratopsians together into a broader group. However, as we know so little about pachycephalosaur origins, it is possible that they might be related to other ornithischians instead. Pachycephalosaur fossils should be present in older rocks, as all their potential relatives lived during the Jurassic, but so far nothing has been found. We desperately need new fossils to reveal how pachycephalosaurs gained their distinctive features and to fit them more snugly into the dinosaur tree of life.

# Thrills and frills
## *Triceratops*

WHEN I WAS GROWING UP, *Triceratops* was my favourite dinosaur. With its huge, ornate skull it looked more like a fantastic heraldic beast than a real, once-living animal. Although I have now worked on many other dinosaurs, I still get a thrill each time I see one.

*Triceratops* is the most famous member of the ornithischian group Ceratopsia, also known as the horned dinosaurs, but it was only one of many different ceratopsian species, which varied considerably in size and shape. Rather confusingly, the earliest-known members of the group, such as *Yinlong* from the Late Jurassic of China, did not have horns. However, even these ancestral ceratopsians have two skull features that are present in all members of the group, which help to unite them. The first of these is the appearance of a new bone called the rostral, not seen in any other dinosaur group, which forms the tip of the snout, giving it a stout, curved, parrot-like beak. The second is a pinched-out, narrow shelf of bone at the back of the skull – the beginnings of what would become the extravagant frill of later ceratopsians. Unlike their later relatives, *Yinlong* and other early ceratopsians were small animals, up to 2 m (6½ ft) in length, and were bipedal with long thick tails. They had deep, powerful jaw bones, with closely packed teeth that sheared past each other precisely, enabling them to chop and slice through the toughest plants. All the earliest ceratopsians lived in East Asia, where they flourished, and they spread to North America (and possibly Europe) during the second half of the Cretaceous. One early ceratopsian, *Psittacosaurus*, is known from thousands of fossils, ranging from tiny hatchlings to adults, that come mainly from China, but also from Mongolia, Russia and Thailand.

During the middle part of the Cretaceous, ceratopsians abandoned their bipedal ancestry and went down on to all

One of the most iconic dinosaurs, *Triceratops* is often shown in combat with *Tyrannosaurus*. Some fossils provide direct evidence of this behaviour with *Tyrannosaurus* tooth marks on *Triceratops* bones.

fours, becoming quadrupeds. At around the same time, the tiny frills seen in their earlier relatives expanded greatly, flaring back over the neck to form huge, shield-like structures. The skulls and jaws became deeper and more powerful, and huge, hook-like beaks developed on the upper and lower jaws for ripping and tugging at vegetation. They also started to increase in overall size. This next stage in ceratopsian evolution is exemplified by animals like *Protoceratops* from the Late Cretaceous of Mongolia and China, which reached up to around 2.5 m (8¼ ft) in length and weighed around 100 kg (220 lb).

The last ceratopsians to appear, and the most iconic, are a subgroup called the ceratopsids, which includes *Triceratops*. In these species, the features established in earlier members of the group were pushed towards extremes. The frills became even larger and more elaborate, with these giant skulls accounting for up to one-quarter of the animal's total body length. The record for the longest skull of any land animal that has ever lived is held by a ceratopsid called *Pentaceratops* (sometimes called *Titanoceratops*), whose head was a remarkable 2.6 m (8½ ft) in length, which is exceeded by only a few species of whale. In addition to increasing in length, additional bony projections adorned the margins of ceratopsian frills, which were absent in their earlier relatives, and some of these were long, curved and spike-like. Many ceratopsids

also developed horns. Famously, *Triceratops* has one short, stout horn on the tip of its snout and two long brow horns, positioned above the eyes and facing forwards: each of these would have been covered in a layer of keratin and could reach over 1 m (3¼ ft) in length. Horn shapes and sizes varied between different ceratopsid species: sometimes they were absent altogether, with thickened bony bosses in their place, as in *Pachyrhinosaurus*; other times, there were three horns; often a single horn above the nose, as in *Centrosaurus*. Variations in frill size, outline, the shapes of the frill margins and the numbers, size and positions of the horns help to define the different ceratopsid species, in ways similar to the differences between the horns and antlers of living antelope and deer. Traditionally, it was thought that ceratopsids used their horns for defence, in classic confrontations with giant theropods like *Tyrannosaurus*. Although scientists still think that the horns could have been useful for fending off predators, they now believe that they probably evolved for an entirely different reason – for contests with each other. The large, showy frills might have been excellent for showing off in contests over mates or social dominance, and it is possible that they were brightly coloured; whereas the horns would have been useful for locking the heads of competing animals together as they engaged in trials of strength. Deer and antelope use their impressive headgear for these reasons today, and it seems reasonable to infer that ceratopsids, which were also large, sociable herbivores, might have done the same. Ceratopsids were the last and largest of the ceratopsians, with *Triceratops* weighing in at up to 10 tonnes (11 tons) and reaching 10 m (33 ft) in length.

Although East Asia was the cradle of ceratopsian evolution, ceratopsids were surprisingly rare in this region. However, they did spectacularly well in North America, where they became some of the most important herbivores in the last Cretaceous ecosystems. A few tantalizing, but scrappy, finds have hinted at the presence of ceratopsians into South America and Australia, but these discoveries are controversial and it appears that the group was restricted to the northern hemisphere.

A single brow horn of *Triceratops*, which was collected from the latest Cretaceous Hell Creek Formation of Wyoming, USA. In life it would have been even longer as the bony core would have been covered with a keratin sheath.

# Mesozoic cows
## *Hypsilophodon*

BY FAR THE MOST ABUNDANT and diverse ornithischians are the ornithopods, a group of herbivores that developed complex, sophisticated chewing methods for dealing with their plant-based diets. In many ways, ornithopods were the dinosaur equivalents of the sheep, cows, antelope and deer with which we are familiar today; many ornithopod species lived in herds, and they would have spent most of their days with heads down, browsing on the ferns, cycads and other low-growing plants that surrounded them during the Jurassic and Cretaceous. At first glance, the most primitive ornithopods, like *Hypsilophodon* from the Early Cretaceous of England, are similar to other early ornithischians: they were small and bipedal, racing around on long, powerful hind legs. They had very long, muscular tails that sometimes accounted for up to two-thirds of the animal's total length, which assisted in balance when running, and they used their small, nimble hands to grapple with vegetation (and maybe each other from time to time). However, even the earliest ornithopods possessed characteristics that were unique to the group, many related to feeding. For example, *Hypsilophodon* and its kin had more precise, efficient jaw movements than other plant-eating dinosaurs, enabled by changes to the back of the skull, the lower jaws and the jaw-closing muscles. These, in turn, allowed the teeth to grind past each other tightly, giving them the ability to chew their meals thoroughly. This was important, because it allowed ornithopods to extract more nutrients from the plants they were eating, as extensive chewing before swallowing speeds up and improves the digestive process.

    The fossil record shows that several different types of ornithopod had evolved by the Late Jurassic, with the remains of many different species being found in North America, Europe, Africa and Asia. To achieve this wide distribution, their ancestors must have appeared earlier, probably in the Middle Jurassic. Oddly, however, there are only a handful of Middle Jurassic ornithopod fossils, and filling this gap in their origins is an important goal for the

This skull is the holotype (name-bearing reference specimen) of *Hypsilophodon*, from the Early Cretaceous rocks of the Isle of Wight, UK. When first found it was thought to be a baby *Iguanodon*, but it was soon realized that it belonged to an earlier branch of ornithopod evolution.

scientists working on this group. Although Late Jurassic ornithopod fossils are fairly common, counting the numbers of different dinosaur remains at fossil sites shows that they were not the most important members of any dinosaur communities at this time – instead, the commonest herbivores, by far, were the long-necked sauropods (and stegosaurs were important in some places too). However, things were set to change – by the Early Cretaceous, a quiet revolution had taken place in the numbers and types of dinosaurs present, especially in the northern hemisphere. Sauropod dinosaurs became gradually less important on the northern landmasses of North America, Europe and Asia, although they remained the commonest plant-eaters in South America, India and Africa. By contrast, in the north, ornithopods began to take the place of sauropods as the main large-bodied herbivores and, as they evolved, they developed a wide variety in their appearance and behaviour.

At this time, another type of ornithopod dramatically increased in importance. In these more advanced ornithopods the features present in animals like *Hypsilophodon* were modified further, increasing feeding efficiency. These dinosaurs were also much larger: *Hypsilophodon* tipped the scales at 20 kg (44 lb), but its heavier relatives could reach 5 tonnes (5½ tons) or more. These larger ornithopods, the iguanodontians, first appear in the Middle Jurassic, but it was during the Early Cretaceous that they came to dominate many dinosaur ecosystems. They exploded in numbers, becoming the most numerous dinosaurs in Europe and North America, and many new species appeared. One of the most famous was *Mantellisaurus*, which lived alongside *Hypsilophodon* in southern England.

*Mantellisaurus* possesses the key features that underpinned iguanodontian success. It was much larger than the earliest ornithopods, which conferred additional protection from predators and allowed it to have a much longer gut for digesting tough plant food. Larger size came at a cost, however: it was no longer as agile as its smaller relatives and would not have been able to rely on speed for defence. Iguanodontians also had unusual hands in comparison to other ornithischians, with several specialized features. The middle three fingers were adapted for weight-bearing, by becoming stronger, stouter and developing large, hoof-like claws. These features, combined with changes to the hips and hind limbs, show that iguanodontians spent most of their time on all fours, although they might have been capable of walking bipedally for short periods while

All known *Hypsilophodon* specimens are small and adult individuals have yet to be found. This skeleton, from a young juvenile, is less than 1 m (3¼ ft) long.

they trotted or ran. Although the hands were used primarily for support, they had other functions. *Mantellisaurus*, and close relatives, had a thumb tipped with a cone-shaped spike. This spike is a modified finger bone and was probably used as a weapon for warding off predators (and perhaps used in fights over food or mates). Finally, the skulls of iguanodontians were longer and lower than those in earlier ornithopods, and were equipped with rows of much larger, tightly packed teeth that form a lengthy, continuous surface for grinding plant food.

Iguanodontians spread around the world, ranging from Antarctica to Siberia, but their fossils are best known from Europe, the USA and eastern Asia. They were immensely successful during the Early Cretaceous, with some species reaching up to 10 m (33 ft) in length. Thanks to the large number of species, some of which are known from multiple complete skeletons their anatomy is among the best understood of all dinosaur groups. For example, a famous site in Belgium, called Bernissart, has yielded the remains of more than twenty complete *Iguanodon* skeletons. Iguanodontians also hold a special place in dinosaur studies as *Iguanodon* was the second dinosaur to be named scientifically, by Gideon Mantell in 1825. However, they dwindled in terms of both numbers of individuals and species during the Late Cretaceous. Their decline might have been linked to the appearance of a new group of even larger ornithopods, which were descendants of animals like *Mantellisaurus*. These new ornithopods were the mighty hadrosaurs.

# Honking duckbills
## *Edmontosaurus*

THE HADROSAURS, OR DUCK-BILLED dinosaurs, were the dominant herbivores in Late Cretaceous North America, where their fossils are amazingly abundant, but they spread worldwide and lived on every continent except Australia. They were the last group of ornithopod dinosaurs to evolve and, in many ways, the most advanced. One of the best-known hadrosaurs, represented by hundreds of skeletons in museum collections, is *Edmontosaurus* from the very latest Cretaceous of the USA and Canada. Given how well-known its skeleton is, *Edmontosaurus* provides a great model for showing off the features that promoted hadrosaur success. Hadrosaurs were, on average, larger animals than their iguanodontian ancestors, and *Edmontosaurus* reached around 12 m (39½ ft) in length and 6 tonnes (6½ tons) in weight (the largest of all hadrosaurs, *Shantungosaurus* from China, reached a colossal 16 m (52½ ft) and 13 tonnes (14¼ tons)). Hadrosaurs are nicknamed 'duckbills' because of the wide, flattened, toothless tips to their snouts,

*Edmontosaurus* was a wide-ranging dinosaur whose remains are abundant all over western North America, from Colorado to Alaska. This distribution suggests that it was capable of thriving in a wide variety of ecological settings.

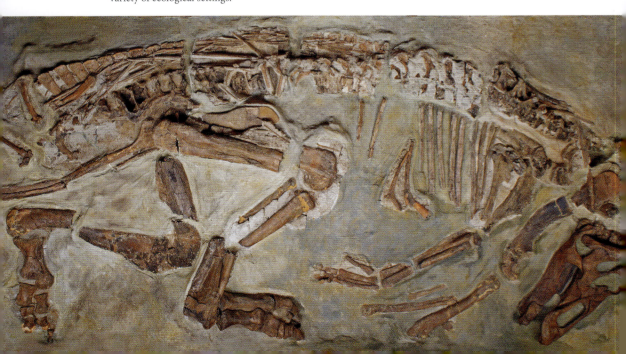

which were covered with sharp horny beaks for cropping tough leaves and stems. Further back in the skull, the jaws are lined with hundreds of teeth. An individual *Edmontosaurus* might have up to 1,000 teeth in its mouth at any one time, and other hadrosaurs had similar numbers. The teeth are small, with diamond-shaped tips, and fit tightly together with no gaps between them, forming a continuous rough pavement for pulverizing food. These complex 'dental batteries' are one of the features that distinguishes hadrosaurs from their earlier iguanodontian cousins.

Hadrosaur hands are surprisingly small and dainty in comparison with the rest of their bodies, and they lack the thumb spikes of iguanodontians. The fingers are tipped with a large, broad hoof and trackways show that hadrosaurs walked on all fours. One of the most striking features of their skeletons are the criss-crossing ribbons of bone running alongside their backbones: these are structures called tendons. Tendons are normally, springy strips of tissue that connect bones and muscles together, but in hadrosaurs some turned into solid bone, making the tail and back much stiffer than normal, which might have helped to support their weight more effectively.

Some exceptional fossils of *Edmontosaurus* show that it had a row of impressive scales running along the middle of its back and a soft fleshy crest on its head, a bit like a cockscomb. These and other features suggest that display was an important feature of hadrosaur biology. Some hadrosaurs are famous for the impressive bony crests arising from the backs of their skulls, such as the long, gracefully curving crest of *Parasaurolophus*, which could reach over 1.5 m (5 ft) in length. These crests were sometimes hollow, with their internal passageways connected to the nostrils, and they could be used to make a variety of deep, honking and bellowing sounds, providing another surprising similarity with geese and ducks.

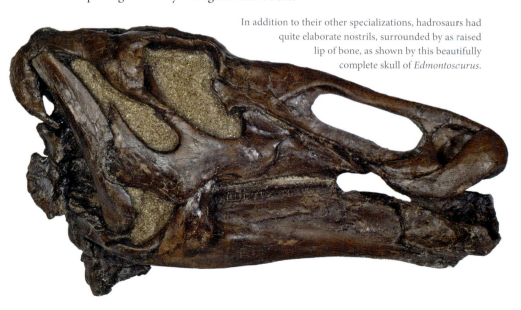

In addition to their other specializations, hadrosaurs had quite elaborate nostrils, surrounded by as raised lip of bone, as shown by this beautifully complete skull of *Edmontoscurus*.

# Dinosaurs go global
## *Pantydraco*

DURING the Late Triassic or Early Jurassic, the commonest dinosaurs were early sauropodomorphs. These animals lived everywhere, from *Glacialisaurus* in Antarctica to *Issi* in Greenland, and they inhabited a range of environments from inhospitable sandy deserts to lush, well-watered floodplains. As a group, they were versatile, some were small, bipedal omnivores, while others were huge, possibly quadrupedal, herbivores. Their dietary flexibility was probably a key factor in allowing them to live in so many different places, especially in areas where food supplies might have varied strongly from season to season.

Basal sauropodomorphs were pioneering dinosaurs and were the first to achieve some of the things that became important in the spread and later success of dinosaurs as a whole. For example, they were the first dinosaurs to live in herds, as shown by mass accumulations of skeletons for species like *Plateosaurus*, during the Late Triassic. This has obvious implications, not only for defence against predators, but also for dinosaur social behaviour. Sauropodomorphs were the first dinosaurs to attain weights greater than 1 tonne (1 ton), a feature that was helpful both in plant feeding, with longer guts for more efficient digestion and greater feeding heights, and defence. In addition, they were innovators in herbivory, which opened a whole new way of life for dinosaurs to exploit, quite different from those of their omnivorous and carnivorous ancestors. One of the feeding adaptations specific to sauropodomorphs was the evolution of their characteristically long necks, which gave access to taller vegetation, a food source that was unavailable to the smaller, shorter-necked herbivores that lived before and alongside them. Finally, the earliest fossil eggs, nests and embryos currently known for any land-living backboned animal belonged to two early sauropodomorphs, *Mussaurus* and *Massospondylus*, which lived in the Early Jurassic. These important remains provide our first glimpses into the changes that occurred during the growth of early dinosaurs from hatchlings to adults, and they allow comparisons with how growth takes place in living relatives, such as crocodiles and birds.

Despite their abundance, early sauropodomorphs went extinct by the end of the Early Jurassic, achieving a much sadder first – becoming the first major group of dinosaurs to vanish from the Earth. We are still unsure of the reasons for their extinction: there were some minor environmental changes at this time, but nothing that would obviously lead to their demise. It is possible that they faced stiff competition from their own descendants – a more specialized group of sauropodomorphs that went on to include the largest land animals of all time, the sauropods.

The skull and neck of the small, probably omnivorous, sauropodomorph *Pantydraco*, from the Late Triassic of Wales. *Pantydraco* lies close to the base of the sauropodomorph family tree, and exhibits the basic 'ground plan' that was modified by later members of the group.

# The first giants
## *Mamenchisaurus*

SAUROPODS INCLUDE SOME of the most iconic dinosaurs, and species like *Brontosaurus* and *Brachiosaurus* are both famous and familiar. Sauropods inherited many of their most characteristic features from their ancestors, the early sauropodomorphs, but pushed these to extremes, creating a stable, instantly recognizable body plan that underpinned their staggering diversity. The first true sauropods appeared in the Early Jurassic and lived alongside their early sauropodomorph relatives in South America and Africa. However, these first sauropods were rare and much smaller than their later relatives. It was only after the extinction of early sauropodomorphs, at the close of the Early Jurassic, that sauropods began to diversify into a wider variety of types, eventually achieving a global distribution.

All sauropods have the same basic body shape, as typified by *Mamenchisaurus* from the Late Jurassic of China. Sauropods abandoned the bipedality seen in their ancestors to become exclusively quadrupedal, a change enabled by modifications to the whole body, but especially the limbs. The forelimb bones became stouter, stronger and longer, and these alterations allowed the limbs to be held directly under the body. The hands lost their grasping functions and the hindlimbs turned into straight, upright columns with massive muscle attachments to the hips and tail. When combined, these changes meant that all the limbs were ideal not only for walking but also for bearing weight. Interestingly, even the very first sauropods, like *Vulcanodon* from the Early Jurassic of Zimbabwe, were relatively large animals in comparison with most early sauropodomorphs, and it seems that sauropods developed quadrupedality at the same time as large size, with the former potentially allowing the evolution of the latter.

Sauropods are unmistakable. It has been said that they look a little like an elephant that has swallowed a python. Their bodies are massive and barrel-shaped, with the sturdy limbs supporting each corner, and extraordinarily long necks and tails sticking out in opposite directions. All sauropods had these incredible necks, which developed in two main ways. First, sauropods added extra bones to the neck: most other dinosaurs had nine or ten neck vertebrae, but sauropods had at least 12. In some cases, sauropods gained these by 'capturing' vertebrae that were already part of the trunk and adding them to the neck (with the consequence that their trunks became a bit shorter at the same time). In other cases, sauropods added brand new vertebrae to the neck, increasing the total number to 15 or more. As well as adding new neck vertebrae, each individual neck vertebra became much longer: some vertebrae were well over 1 m (3¼ ft) in length. This combination allowed the evolution of some truly spectacular necks. *Mamenchisaurus* currently holds the record for having the longest neck of any animal that has ever lived. It had 18 neck vertebrae and one species, *Mamenchisaurus sinocanadorum*, had a neck that reached a staggering 16 m (52½ ft) in length. These super-long necks allowed sauropods access to vegetation over wide horizontal and vertical ranges. They probably had other uses too: the necks were so prominent that they might have played a role in visual signalling and communicating with other sauropods. They were also very muscular and could have been used for fighting other sauropods or for taking swipes at predators.

Sauropods were dedicated herbivores and their skulls and teeth reflect this. The skulls became relatively taller (top to bottom) and shorter (front to back) than those of early sauropodomorphs, which gave these dinosaurs more powerful bites, and their teeth became larger and met each other more precisely as the jaws closed, which allowed more efficient slicing of plant food. Really large sauropods would have eaten several hundred kilogrammes of food every day to fuel their huge, multi-tonne bodies.

The vertebrae of the neck, back and hip regions, and very often the part of the tail closest to the body, are hollowed out in a variety of different ways. The sides of the vertebrae often bear deep depressions, with further channels and holes within them that lead into the interior of the bone. These hollows and holes reflect the presence of air sacs that were present in the living animal. The air sacs are balloon-like extensions that are connected to the lungs and were an important part of the sauropod breathing apparatus, providing extra space in the system that allowed them to move air up and down their very long necks. The air sacs spread throughout the whole body – including the bones themselves. In addition to being critical for sauropod breathing, the air sacs also made the skeleton lighter than expected, helping these giant animals to save some weight. The tops of the vertebrae, which provided the areas for back muscle attachments, are also hollowed out into a series of complicated, elaborate struts.

Following their first appearance in the Early Jurassic, sauropods spread around the world and their remains have been found on all continents. During the Middle and Late Jurassic they evolved into dozens of new species and became the most important herbivores on land. The early sauropods started to set new records in terms of body mass, with many of these animals weighing 10 tonnes (11 tons) or more. But in the Late Jurassic and Cretaceous, sauropods started to achieve even more impressive body masses, sometimes exceeding 50 tonnes (55 tons), a feat rivalled only by whales in the history of life, and by no other land-living animals.

OVERLEAF: *Mamenchisaurus sinocanadorum* was just one member of a group of exceptionally long-necked sauropods, known as mamenchisaurids. All mamenchisaurids, whose remains are known mainly from China, seem to have had necks that were more than three times the length of the trunk.

# Whip-tailed sauropods
## *Diplodocus*

DURING THE EARLY AND Middle Jurassic, sauropods underwent a burst of evolutionary experimentation. Several different groups appeared: most of these flourished briefly but vanished without descendants. However, two of the groups originating at this time survived, and went on to produce the hundreds of sauropod species seen in Late Jurassic and Cretaceous ecosystems. One of these groups was the Diplodocoidea, named after its most famous member, *Diplodocus*, from the Late Jurassic of the USA. Although they retained the same basic design as other sauropods, diplodocoids had many unique features of their own.

Perhaps the most striking of these relate to the skull. When viewed from the front or the top, the snout is very wide and has a rectangular outline, which contrasts with the narrower, more rounded snouts of other sauropods. This gives *Diplodocus* its rather goofy, charming 'grin' but also, and more importantly, increased the breadth of the mouth and the efficiency with which it could gather food. These changes to the snout were accompanied by radical transformations of the teeth. In *Diplodocus* and its relatives, the teeth are long, narrow, pencil-like structures, which are totally different from the broader, chomping teeth of early sauropods. These pencil-like teeth were useless for chewing, because they did not meet precisely or grind past each other, but they were ideal for use as rakes or combs to strip branches and stems of their leaves. Diplodocoid noses are also

Some scientists have suggested that the long, thin part of the tail in *Diplodocus* could have moved at supersonic speeds, making a cracking sound like that of a bullwhip. Others think it could be flicked hard for defence or used to communicate with other herd members through touch.

rather odd. In all other dinosaurs, the nostrils are near the front of the skull, with one each side, but in *Diplodocus* and its relatives, these holes have merged to form one large nostril on the middle of the skull, and this opening has moved backwards to lie above and between the eyes. Scientists are not sure why diplodocoid noses are so unusual, and the evolution and function of this feature remains mysterious.

We always think of sauropods as giants and it is true that diplodocoids grew to impressive sizes, with *Supersaurus* reaching an astonishing 34 m (111½ ft) in length. However, although diplodocoids are contenders for being the longest sauropods, they had a rather slender build in comparison with the other sauropod groups, so they were not the heaviest. Some diplodocoids, such as *Dicraeosaurus* from Tanzania, were actually rather small by sauropod standards, weighing in at 'only' 5 tonnes (5½ tons). Looking at the rest of the skeleton, diplodocoids often had shorter forelimbs than other sauropods, so their chest was closer to the ground, which might have allowed them to feed at lower levels more easily. In addition, many had exceptionally long tails with 70 or more vertebrae, with the final few bones of the tail forming a slender, whip-like structure, that might have been used in defence or communication.

# Big-nosed behemoths
## *Giraffatitan*

GIRAFFATITAN TOWERED OVER THE coastal plains of what is now Tanzania during the Late Jurassic period. One of the largest dinosaurs of its time, this gargantuan sauropod – with its exceptionally long, elegant neck that was composed of 15 elongate vertebrae – had a vertical reach of 12 m (39½ ft), tall enough to peer into a fourth-storey window. Although *Giraffatitan* did not attain the extreme body lengths seen in some other sauropods, it still measured an impressive 23 m (75½ ft) from nose-to-tail. What is more, it was very heavily built, and much stockier than the relatively slimline *Diplodocus* – the bulkiest *Giraffatitan* individuals might have reached weights of 40 tonnes (44 tons) or more.

The short, boxy head was equipped with chisel-like teeth, which were in constant use while gathering the vast quantities of vegetation needed to fuel its hulking frame. The front of the skull was stretched to form a broad snout, but perhaps its oddest features were the enormous nostrils, framed by a delicate, arched strut of bone. The bony frames of the nostrils were much larger than *Giraffatitan*'s eye sockets, although we are not sure why they should have reached such prodigious sizes (the fleshy openings of the nostrils, however, would have been much smaller). However, *Giraffatitan* was not alone in possessing such an impressive nose: a whole group of sauropods share this unusual feature, and this, together with a number of other characteristics spread throughout the skeleton, suggests that they were all close evolutionary relatives. This sauropod group is named Macronaria – literally 'big noses' – and includes the majority of Cretaceous sauropods as well as their earlier Jurassic relatives.

*Giraffatitan* and other macronarians, such as *Camarasaurus* from the Late Jurassic of the USA, can be distinguished from their diplodocoid cousins by features other than their noses. All macronarians share many features of the vertebrae, limbs and hips that help to define them as a group. Perhaps the most obvious of these are the forelimbs, which become enlarged relative to their hind limbs, so that the chest and shoulders were held higher than those of other sauropods. This reached an extreme in *Giraffatitan* and *Brachiosaurus*, where the forelimbs were longer than the hind limbs, a feature that

OPPOSITE: The famous mounted skeleton of *Giraffatitan*, which is on public display in the Museum for Nature, Berlin, Germany, is the largest real dinosaur skeleton on display anywhere in the world.

is unique among dinosaurs, and that resulted in the animal's back sloping downwards from its shoulders to its hips.

The first macronarians appeared at around the same time as the first diplodocoids and both groups thrived, with particularly diverse and mixed sauropod faunas emerging in the Late Jurassic. This period is regarded by many palaeontologists as the peak of sauropod evolution, due to the variety of new sauropod species appearing, and their rapid spread around the world. Their wide geographic range was matched by a marked increase in their ecological importance, with sauropods becoming the most important herbivores in most Jurassic ecosystems, a role that continued in many parts of the world through to the end of the dinosaur era.

Macronaria includes the largest animals ever to live on land. These behemoths belong to a macronarian sub-group with the rather appropriate name Titanosauria – literally the 'titan reptiles'. Titanosaurs were the most successful sauropods, with the greatest number of species. They spread all over the world, reaching every continent, and they were the most abundant sauropods during the Cretaceous Period (diplodocoids were still present during the Early and mid-Cretaceous but in small numbers relative to their Late Jurassic peak). Titanosaurs possessed many distinctive features. For example, some had plates of bone (osteoderms) embedded in their skin, which could be up to 60 cm (2 ft) long. These might have been useful for defence, or for storing minerals, like calcium and phosphorus, for later use in growth or reproduction. They had unusually wide hips and very stout, chunky limbs, and these modifications provided extra support and gave them a unique 'wide-gauged' stance, that produced characteristically broad trackways.

During the course of their evolution, titanosaurs showed changes in body size to a greater degree than any other sauropod group. Many titanosaurs were comparable to 'average-sized' sauropods, such as *Camarasaurus* and *Diplodocus*. Others surpassed the already impressive dimensions of their close cousin, *Giraffatitan*: some South American titanosaurs reached weights of at least 60 tonnes (66 tons) – and perhaps 70 tonnes (77 tons) in the case of the poorly known, but epic, *Argentinosaurus*. However, not all titanosaurs were giants – some of the smallest known adult sauropods belonged to this group, such as the diminutive *Magyarosaurus* from the very latest Cretaceous of Romania. This animal, which was tiny by titanosaur standards, reached a maximum length of 'only' 6 m (19¾ ft) and probably weighed less than 1 tonne (1 ton). Although this sounds impressive, and is certainly bigger than the majority of living land mammals, it is worth bearing in mind that the average weight of a fully grown sauropod was 20 tonnes (22 tons) or more. Instead, *Magyarosaurus* was closer in weight to a very large bull and reached only a fraction of the weight of an elephant. Indeed, the thigh bone of *Magyarosaurus* is not much longer than that of a human male – most of its 6 m (19¾ ft) length was occupied by its lightweight neck and tail. This small titanosaur probably owes its unusual size to the fact it lived on what, at the time, were islands – and these islands did not have the resources to support herds of huge plant-eaters. It seems likely that these dinosaurs shrank over evolutionary time, as they are descended from much larger ancestors, and this phenomenon was an adaptation to deal with the constraints of island life.

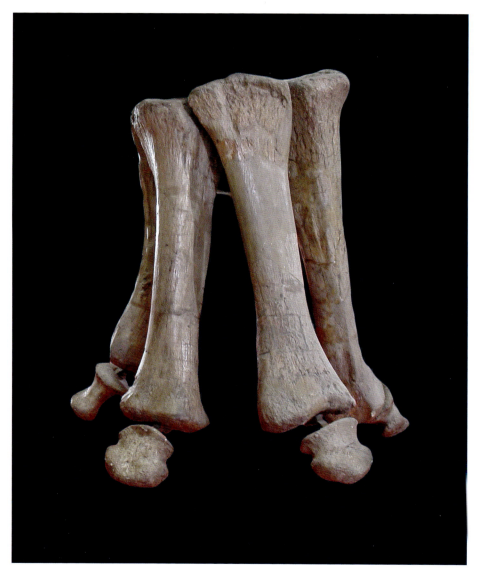

Sauropod hands, including those of *Giraffatitan*, were long, slender and tube-like in build, adding extra length to the forelimbs.

# Crested hunters
## *Dilophosaurus*

THEROPODS DIVERSIFIED IN THE Late Triassic, with new species appearing all over the world. However – with a few exceptions – these early hunters varied little in size, shape and behaviour, with most looking superficially similar to the fleet-footed *Coelophysis*. They foraged alongside other, more fearsome, carnivorous reptiles such as phytosaurs and rauisuchians, and were probably prey themselves on a regular basis. However, perhaps surprisingly, the upheavals of the end-Triassic mass extinction would come to their aid, by removing these apex predators from the ecological stage forever. The absence of these competitors provided theropods with new opportunities, which they exploited rapidly, resulting in their inexorable rise to the top of the food chain.

Jurassic theropods were, on average, much larger than their Triassic cousins. One of the best-known Early Jurassic hunters is *Dilophosaurus*, from the southwestern USA. *Dilophosaurus* is most famous for its appearance in the film *Jurassic Park*, although, in reality, it was quite different from its cinematic avatar. For a start, it was much, much larger – with adults reaching 7 m (23 ft) in length, and it would have been an intimidating foe. The film's *Dilophosaurus* had an elaborate neck frill, which it flourished when making its lethal attack; it also spat venom at the hapless victim. However, although these behaviours added excitement to the story, and are based on the biology of living lizards and snakes, both are pure fantasy. No dinosaur had an extendable neck frill and, although one or two might have been venomous (on the basis of a few teeth with a possible venom delivery groove), there is no evidence for this in *Dilophosaurus*.

Nevertheless, one bizarre feature of *Dilophosaurus* is definitely real. It sports two large, fan-shaped bony crests on top of its snout, which extend backwards from the nostrils and end just in front of the eyes. The surfaces of the crests are grooved and ridged, indicating that they were covered by a horny sheath made of the tough protein keratin. It is possible that this keratinous covering was brightly coloured and it

This partially complete skull of *Dilophosaurus* shows the thin, delicate fan-like crest that gave the animal its name.

seems likely that these structures were used for display, suggesting that *Dilophosaurus* was capable of some complex social behaviours. Interestingly, many other Jurassic theropods also developed crests. Fan-shaped crests like those of *Dilophosaurus* were present in the Chinese theropods *Sinosaurus* and *Monolophosaurus* (although, in the latter, only a single crest was present on the midline of the skull). A bizarre wave-like crest was present on the forehead of *Cryolophosaurus* from Antarctica, and the Late Jurassic *Ceratosaurus* bears a horn-like projection on its nose. All these features suggest a burst of anatomical evolution as Jurassic theropods experimented with different lifestyles. However, for reasons we do not understand, this flirtation with impressive headgear did not catch on – although some later theropods, such as *Oviraptor*, do boast impressive crests, the majority of theropods dispensed with elaborate bony ornaments. Recent discoveries have shown that many Cretaceous theropods found other means of display: they were covered in feathers.

# Southern predators
## *Carnotaurus*

ABELISAUROIDS WERE A DIVERSE but mysterious group of carnivores that flourished on the southern continents. Although one early species (*Eoabelisaurus*) is known from the Lower Jurassic rocks of Patagonia, almost all of their remains come from the Cretaceous Period, and there are frustrating gaps in their fossil record. Similarly, most species are represented by incomplete skeletons, but it is clear that abelisauroids lacked the features present in other, more advanced, theropods. They had a long, separate evolutionary history, which paralleled that of the other, largely northern, theropod groups. This culminated in the appearance of two major abelisauroid sub-groups. One was Noasauridae – agile small-bodied hunters, like the bizarre, buck-toothed *Masiakasaurus* from Madagascar. The other sub-group, Abelisauridae, followed a different trajectory, increasing in size to become apex predators in many Gondwanan ecosystems.

One of the most complete, and striking, abelisaurids is surely *Carnotaurus* – literally 'the meat-eating bull' – a medium-sized, lightly built hunter that roamed Argentina during the middle of the Cretaceous. Even from a distance, you would have realized that this 8 m (26¼ ft) long predator differed somewhat from the more familiar tyrannosaurs that lived at the same time in North America. For a start, the head of *Carnotaurus* (and other abelisaurids) is unusual. Most theropods had snouts with long, low profiles and, when viewed from above, the skull was narrow from side-to-side. By contrast, *Carnotaurus* has a snub-nosed, tall, wide and boxy skull. These proportional differences suggest that abelisaurids were dealing with prey in quite different ways from other theropods – short, deep skulls imply a strong, gripping bite, and it is possible that *Carnotaurus* used its jaws to grab and hold on to prey for extended periods, rather than making quick, slashing bites. This idea is backed up by the structure of the neck, which was stocky and powerfully muscled.

*Carnotaurus* gets its name from its most distinctive feature: a pair of short, cone-shaped horns positioned above the eye sockets, which taper to a point and project upwards and slightly outwards. The surfaces of the horns, and the other skull bones, are textured and rough, suggesting that they were covered with keratin. *Carnotaurus* is the only horned theropod, and it is likely that these impressive ornaments were used in contests over mates or territory. Patches of lizard-like scaly skin are preserved with the skeleton and it is unclear whether these primitive theropods had feathers.

Like the distantly related tyrannosaurs, the arms and hands of abelisaurids became greatly reduced, and they relied on their jaws to apprehend and dispatch prey. Unlike tyrannosaurs, however, *Carnotaurus* had four-fingered hands, although the fingers were tiny and immobile, and the hands were smaller than those of a human child. Their function remains unknown: display is possible, but they might have been evolutionary 'leftovers', remnants of an earlier feeding strategy. It is tempting to speculate that they would have disappeared entirely if the group had not become extinct.

The deep, boxy skull of *Carnotaurus* suggests that it was capable of forceful bites. The stout, triangular horns jut upwards from above the eye sockets.

# Aquatic dinosaurs
## *Spinosaurus*

ARISTOCRATIC German palaeontologist Ernst Stromer von Reichenbach had an exciting theory. He believed that humans evolved in Africa, an idea that is now well-established, but for which there was little evidence at the start of the twentieth century. Stromer decided to test his hypothesis by searching for fossils in the desert of Egypt, mounting several expeditions to the Bahariya Oasis, with the first commencing in 1910. Although his search for human ancestors failed, he did succeed in finding something equally spectacular: the tantalizing remains of a huge, bizarre theropod – *Spinosaurus*.

Among the fossils discovered by Stromer was a long, curved lower jawbone bearing large conical, spike-like teeth. These teeth lacked the steak-knife serrations seen on other theropod teeth, suggesting that they were more likely to be used for impaling than slicing. Even more intriguing were the vertebrae, each of which sported an exceptionally long, ribbon-like process, many times the height of the spool-shaped vertebral body. Stromer realized that these processes must have supported a conspicuous sail-like structure that would have extended along the animal's back, from the shoulders to the hips, a feature captured in the name he coined in 1915 (literally, 'spine lizard'). Continued searches in

the Cretaceous rocks of Bahariya brought more bones to light, but complete skeletons remained elusive, limiting Stromer's ability to say more about *Spinosaurus*.

The *Spinosaurus* fossils were brought back to Germany and housed in a museum in Munich. As the century progressed, and war broke out in Europe, the story took a tragic turn. During World War II an RAF bombing raid, on 24 April 1944, reduced parts of the museum to rubble, destroying many of the fossils within, including those of *Spinosaurus*. No other examples were known and for many years after this event Stromer's drawings, photographs and written descriptions were the only evidence that this dinosaur ever existed. It had, in effect, suffered two extinctions. However, palaeontologists working in other parts of north Africa, particularly Morocco, started to unearth new *Spinosaurus* specimens as they prospected previously unexplored Cretaceous fossil beds. Some of these discoveries were more complete than those that had been available to Stromer.

The new fossils not only confirmed many of Stromer's original observations but added valuable new details. In particular, we now know that *Spinosaurus* had an exceptionally long, narrow, crocodile-like snout. *Spinosaurus* also had unusual proportions, with arms that were relatively longer than those in other theropods, and quite powerfully muscled. The possession of these elongate arms has suggested to some that *Spinosaurus* walked on all fours, and it was perhaps the only theropod to accomplish this feat. Although you might think that the crocodile-like skull and quadrupedal stance are unusual enough, it has been proposed that *Spinosaurus* was also the only fully aquatic dinosaur, using its long, deep tail for propulsion through the water. Although the debate

A recent reconstruction of *Spinosaurus* based on newly discovered, more complete remains. The tail is much deeper than those of other theropods and its overall body shape deviates from their usual body plan.

about its swimming abilities divides palaeontologists, most agree that *Spinosaurus* and its relatives were much more reliant on water than other theropods. This is supported by several lines of evidence. First, the rocks yielding *Spinosaurus* remains were laid down in coastal environments. Second, analyses of *Spinosaurus* teeth show that their chemical composition is more similar to that of aquatic animals, like crocodiles, than to those of land-living theropods. Finally, the crocodile-like skull and teeth seem ideally suited for snapping up fish. Indeed, one close relative, *Baryonyx* from southeastern England, has fish scales preserved in its stomach contents, the remains of its last meal.

Despite these new discoveries, the function of the famous sail remains enigmatic. Sails are unusual among vertebrates but are known from a handful of other dinosaurs, a few lizards and several ancient mammal relatives, most famously the carnivorous *Dimetrodon*. It seems most likely that the sail was used for display, as in lizards today – an elaborate signal for advertising information on health, age or status with other individuals. Other ideas for sail function have been proposed, however: perhaps it was an elaborate radiator for helping the animal heat up and cool down. Alternatively, the lengthy spines might have supported a hump-like area of fat storage.

As more *Spinosaurus* specimens came to light they provided additional information on the true size of this animal. Careful measurements and comparisons with other theropods have indicated that some *Spinosaurus* adults reached prodigious sizes – with maximum lengths up to 14 m (46 ft) and a weight of maybe 7 tonnes (7¾ tons). This would make *Spinosaurus* comparable to *Tyrannosaurus rex*, crowning it as one of the largest theropod dinosaurs. However, more complete *Spinosaurus* skeletons (and some other heavyweight species like *Giganotosaurus*) are needed to check these estimates and provide a definitive answer to this question.

So far, all spinosaur fossils have been found in Cretaceous rocks. However, there are likely to have been earlier, so far undiscovered, spinosaur species, because their closest relatives are the megalosaurs that lived during the Middle and Late Jurassic. Although their fossils are not particularly common, and most spinosaur species are known from only a single individual, their remains have now been found all over the world, from *Irritator* in South America, to *Baryonyx* and others in Europe, and *Ichthyovenator* in Thailand. There is even a suggestion of an Australian spinosaur, although this is based on a single neck bone whose identification is controversial. So far, the only major landmass lacking spinosaur fossils is North America. Spinosaurs were most abundant during the early and middle part of the Cretaceous but, despite their many unique adaptations, they disappeared long before the close of the Mesozoic Era.

OPPOSITE: Palaeontologists currently disagree over the swimming ability of *Spinosaurus*. Some think it was an adept swimmer that spent its time cruising through open water, whereas others think it stalked in the shallows.

# Jurassic hunters
## *Allosaurus*

FOSSILS OF THE QUINTESSENTIAL Late Jurassic predator, *Allosaurus*, have been found in dozens of quarries throughout the famous Morrison Formation of the western USA. *Allosaurus* bones have also been found in beds of the same age in Portugal, making it one of a handful of dinosaurs with a truly trans-continental distribution.

Unusually for a theropod, *Allosaurus* is represented by abundant remains, so palaeontologists know a lot about its appearance and lifestyle. One famous site, the Cleveland-Lloyd Dinosaur Quarry in Utah, has yielded the remains of 44 *Allosaurus* individuals, a staggering number, making this one of the richest theropod graveyards in the world (surpassed only by the hundreds of *Coelophysis* skeletons found at the Late Triassic Ghost Ranch site in New Mexico). Top predators are usually rare in any ecosystem – you need a lot of prey to sustain even a small population of large carnivores (there are a lot more wildebeest than lions, for example). As a result, this glut of *Allosaurus* skeletons is puzzling. Something must have drawn them to this particular spot, but there is disagreement over what that could be. It is possible that the area held a vital water source during a drought, which attracted an abundance of prey, although there are far fewer herbivorous dinosaur remains in the quarry than *Allosaurus* bones. Alternatively, the ground might have been soft and sticky, forming a natural trap – a little like quicksand – and waves of hungry *Allosaurus* might have been tempted by mired herbivores unable to escape, but then ended up becoming trapped themselves.

As many *Allosaurus* specimens are so complete, they have been used for pioneering studies of theropod biology. For instance, a beautifully well-preserved *Allosaurus* skull was the first dinosaur specimen to be subjected to a computer-based modelling technique called Finite Element Analysis – a method used by engineers to investigate the performance of complex structures. Computed tomography scans were used to make exact virtual models of its structure and these were subjected to possible feeding loads by the computer. This study found that *Allosaurus* did not close its jaws particularly hard, but that it was capable of fast, slashing bites, almost like striking its prey with a hatchet, so that the unfortunate victim probably died of massive blood loss.

As its remains are so abundant, *Allosaurus* is often used as a reference for interpreting the anatomy of other less complete theropods.

A large predator for its time, *Allosaurus* reached 9 m (29½ ft) in length and was equipped with three-fingered grasping hands, which ended in deadly curved claws. Many similar, closely related theropods shared this body plan and were apex predators in Late Jurassic and Early Cretaceous ecosystems all over the world. These included the similarly sized *Sinraptor*, from the Late Jurassic of China, and the enormous *Carcharodontosaurus*, from the mid-Cretaceous of north Africa (reaching 12 m (39½ ft) in length). This group of hunters, called allosauroids, thrived until midway through the Cretaceous, at which point they dwindled in numbers and other theropods evolved to take their place.

# Feathered theropods
## *Sinosauropteryx*

BY FAR THE MOST VARIED GROUP of theropods is the coelurosaurs. These animals took the basic theropod ground-plan established in the Triassic and stretched it in many different directions. Although the earliest members of the group were small, agile carnivores, they went on to diversify spectacularly, eventually giving rise to gigantic super-predators, bizarre insectivores, crested herbivores and – the most successful dinosaur group of all – birds.

Coelurosaurs differ from other theropods in numerous ways, all of which can be seen in *Sinosauropteryx*, one the best-known members of the group from the Early Cretaceous of China. For a start, they were smart: coelurosaur brains are much larger (relative to their overall body size) than those of other similarly sized theropods. Their hands are further elongated, making them even better at grasping, and their feet are narrow and slender. The ends of their tails are stiffened by the presence of numerous overlapping rod-like bony struts, which rendered them inflexible, a trait that might have helped stabilize the rest of the body during sharp turns when running. However, by far the most innovative characteristic of coelurosaurs concerns their skin: they possessed an entirely new type of skin covering – feathers. *Sinosauropteryx* was the first non-bird dinosaur to be discovered showing unequivocal evidence of feather-like structures, which extended all over its body, and the presence of feathers has since been confirmed in every coelurosaur group.

Although all group members possessed these features, or modified versions thereof, many quite closely related coelurosaur sub-groups looked very different from each other and explored radically divergent ways of life. For example, other than birds, the most famous coelurosaurs are, undoubtedly, the tyrannosaurs. The first tyrannosaurs, like *Proceratosaurus* from the Middle Jurassic of Gloucestershire, England, were small predators, that were not too dissimilar to other theropods. However, during the Cretaceous Period tyrannosaurs evolved to reach huge body sizes, up to 12.5 m (41 ft) long in the case of *Tyrannosaurus*, and developed huge, powerfully-built skulls, with jaws that could deliver bone-crushing bites. Indeed, *Tyrannosaurus* is thought to have had the strongest bite force of any land animal that ever existed, calculated as three

OPPOSITE: The discovery of *Sinosauropteryx* was a sensation, proving that many non-bird dinosaurs had feathers.

times greater than that of a lion. Fossils of early tyrannosaurs from China show that these animals were feathered for at least part of their evolutionary history, but it is still debated if the very largest members of the group kept feathers or had entirely scaly skins (fossils of *Tyrannosaurus* lack feathers, so far, but do have some scales). Similarly, early tyrannosaurs retained three-fingered grasping hands and long arms, but through time these shrank in size and the hands lost a finger each, to produce the famously stumpy arms seen in *Tyrannosaurus* and its closest relatives.

Contrasting sharply with these hyper-predators, the ornithomimosaurs (or 'ostrich-mimicking dinosaurs'), such as *Struthiomimus* from the Late Cretaceous of North America, were graceful, slender sprinters, with small heads set on slender, elegant necks and exceptionally lengthy hind legs that gave them long strides and super speeds. Simulations of ornithomimosaur running ability have suggested that some could reach speeds of up to 80 kmph (50 mph), comparable to a racehorse! Although a couple of ornithomimosaurs reached gigantic size (such as the bizarre *Deinocheirus* from Mongolia, which reached 6.5 tonnes (7¼ tons) in weight) most were much smaller animals with lengths of 3–5 m (9¾–16½ ft) and weighing in at a modest 150–300 kg (330–660 lb), a very different strategy from their tyrannosaur cousins. They also differed from tyrannosaurs, and the majority of other theropods, in losing their teeth. Although the earliest ornithomimosaurs were toothed, later group members lost all of their teeth, and these were replaced by a horny beak that lined the jaws, similar to that of a turtle. As a result, we are not entirely sure what ornithomimosaurs were eating, but it seems most likely that they were herbivores (or perhaps omnivores). We need more fossilized gut contents to be totally sure.

Joining ornithomimosaurs at the Cretaceous 'salad bar' were two more highly modified groups of herbivorous (or omnivorous) coelurosaurs: the oviraptorosaurs (like *Oviraptor* from Mongolia) and the therizinosaurs (including *Erlikosaurus*, also from Mongolia). Most oviraptorosaurs were toothless and beaked, like ornithomimosaurs, but unlike their relatives the skull was shorter and boxier in build, and the tip of the snout was elaborated into a more beak-like shape. They also sported large, complex, helmet-like bony crests on their heads, which seem ideally suited for display. The oviraptorosaur *Citipati* occupies a special place in the history of dinosaur studies, because the first nests to be discovered, in the Gobi of Mongolia, belonged to this animal, showing definitively that dinosaurs reproduced by egg-laying. Close relatives of oviraptorosaurids, the therizinosaurs, went about herbivory in a very different way – elongating the snout and having jaws lined with small, leaf-shaped teeth that were ideal for puncturing and slicing shoots and leaves. Therizinosaurs also had unusual hands, with huge, curved, scythe-like claws. Despite looking fearsome, these claws were probably used mainly for hooking vegetation and pulling it to the mouth during feeding, although they might have played a role in warding off predators. This ability might have been handy as, like many other herbivores, therizinosaurs had large barrel-shaped bodies to house their long digestive tracts, and they were neither speedy nor agile.

Although these coelurosaur groups were quite different in appearance, all of them show new anatomical features that were to become much more important in the origin of the last major dinosaur group we will encounter: the birds.

A close up on the head of *Sinosauropteryx*, showing the halo of dark, fuzzy filaments running along the crest of its head and back along the neck.

# The terrible claw
## *Deinonychus*

FOR DECADES, THE ORIGIN of birds confounded zoologists. Although known to be related to reptiles, their exact ancestry remained obscure. In 1870 the prominent Victorian biologist and palaeontologist Thomas Henry Huxley proposed a radical idea. Citing numerous similarities in their hindlimb and hip structure, Huxley suggested that birds were, in fact, the descendants of dinosaurs. However, his suggestion was not widely accepted: other specialists remained unconvinced, as there were large anatomical gaps between modern birds and those dinosaurs known to nineteenth

and early twentieth century science. Nevertheless, although Huxley's dinosaur–bird hypothesis was rejected by many, there were few other serious proposals, with the exceptions of vague allusions to as-yet unknown Triassic ancestors or potential links with crocodiles. The debate rumbled on without real progress until a stunning breakthrough was made in the 1960s, which reignited the discussion and led directly to our current understanding of bird origins.

Working in the badlands of Montana, John Ostrom, an American palaeontologist, and his team were prospecting the Cloverly Formation, a series of rocks laid down by ancient rivers during the middle part of the Cretaceous Period. Over the course of several expeditions, they recovered over 1,000 bones of a new predatory dinosaur. As Ostrom pored over these remains back in his laboratory, he painstakingly reconstructed the skeleton. One particularly striking specimen was a complete foot, whose second toe was tipped with an enormous strongly curved claw. Ostrom noticed that the joints within the second toe were exceptionally flexible and he realized not only that the claw would normally have been held clear of the ground, but also that it could be rotated through a wide arc. He went on to suggest that this impressive weaponry was employed during hunting, perhaps for slashing the soft undersides and flanks of the unfortunate prey. This 'killing claw' gave the animal its scientific name – *Deinonychus*, or 'terrible claw'.

*Deinonychus* sparked a revolution in dinosaur studies. It showed that at least some dinosaurs were dynamic, fast-moving animals with high metabolic rates, shattering earlier ideas that dinosaurs were sluggish and cold-blooded.

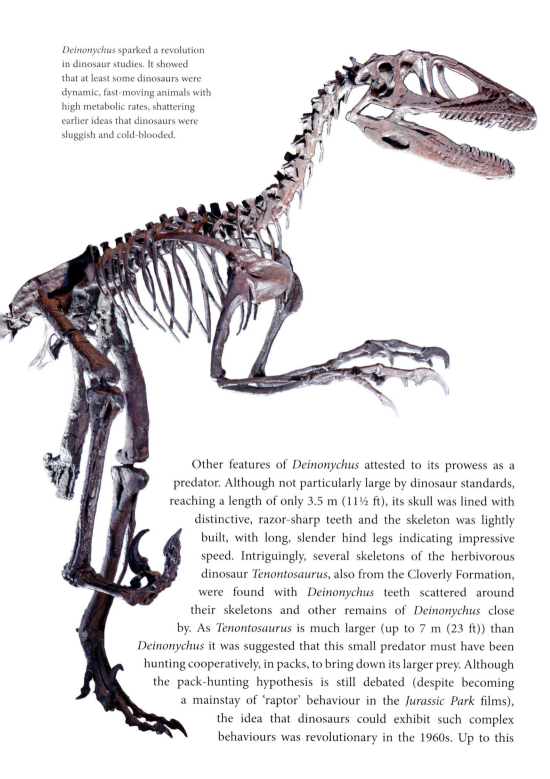

Other features of *Deinonychus* attested to its prowess as a predator. Although not particularly large by dinosaur standards, reaching a length of only 3.5 m (11½ ft), its skull was lined with distinctive, razor-sharp teeth and the skeleton was lightly built, with long, slender hind legs indicating impressive speed. Intriguingly, several skeletons of the herbivorous dinosaur *Tenontosaurus*, also from the Cloverly Formation, were found with *Deinonychus* teeth scattered around their skeletons and other remains of *Deinonychus* close by. As *Tenontosaurus* is much larger (up to 7 m (23 ft)) than *Deinonychus* it was suggested that this small predator must have been hunting cooperatively, in packs, to bring down its larger prey. Although the pack-hunting hypothesis is still debated (despite becoming a mainstay of 'raptor' behaviour in the *Jurassic Park* films), the idea that dinosaurs could exhibit such complex behaviours was revolutionary in the 1960s. Up to this point, they were regarded as incapable of intricate social interactions. In addition, the lithe, slender build of *Deinonychus* indicated that it was able to sustain high physical activity levels, dispelling the widely held notion that dinosaurs were sluggish, scaled-up versions of modern reptiles. It is fair to say that Ostrom's work on *Deinonychus* kickstarted all modern work on dinosaurs, showing that they were not merely slow-moving giants, but vibrant, dynamic animals.

*Deinonychus* had yet another major impact. As he prepared his scientific description of the remains, Ostrom noticed that his new dinosaur, as well as other theropods, shared a surprising number of features with birds. The hands, arms, hips and feet of *Deinonychus* were particularly bird-like. For example, in *Deinonychus* the hand is very long, with slender elongate fingers, and accounts for nearly 40 per cent of forelimb length. This is notable because birds also have long hands, with their fused fingers forming the leading edge of the wings. Zooming in further, *Deinonychus* hands possess a small, bean-shaped bone in the wrist, which has the unwieldy technical name of 'the semilunate carpal'. At first sight, it looks rather boring and unremarkable, but it turns out to be very important. All land-living vertebrates have a series of small, squarish, block-like carpal bones in the wrist, which are important in controlling hand movements and bearing weight. However, birds have a special curved carpal bone, with a more crescentic shape (semilunate means 'half moon'), which gives the hand greater freedom to rotate against the rest of the arm, a feature that is essential in flapping flight. It seems that *Deinonychus* had an early version of this specialized wrist bone, which allowed it to fold its hands back against its body, just as birds do when walking or at rest.

Changes to the shoulder were also bird-like, again allowing more rotation of the joint than seen in other land-living animals. Further along the body, the hips were unusual, with the front hip bone, the pubis, pointing backwards as in birds, rather than forwards as in the majority of other theropods. Although many other features of *Deinonychus* show that it could not fly, the sheer number of bird-like features is startling. Taken together they convinced Ostrom that Huxley had been right all along, and that birds were the direct descendants of small theropod dinosaurs.

Since this ground breaking work, many other small theropods have been studied, closing the gap between birds and dinosaurs even further, so much so that palaeontologists now regard birds as an integral part of the dinosaur family tree. When you see the ducks in the park, or the sparrows on your window ledge, you are looking at a dinosaur.

The famous foot of *Deinonychus*, with its huge 'killing claw' shown raised clear of the ground.

# Taking flight
## *Archaeopteryx*

IN A SPECIALLY BUILT bullet-proof display case at the Natural History Museum, London, you'll see two buff-coloured slabs of limestone. From a distance they are rather nondescript but look closer and you will see, buried within these slabs, delicate, translucent bones, parts of a fragile, splayed out skeleton. The bigger surprise is what surrounds these bones: clear impressions of feathers. This is *Archaeopteryx*, which many palaeontologists consider to be the earliest-known true bird. Discovered in the Late Jurassic rocks of Bavaria in 1861, it achieved celebrity status. Its little skeleton possesses a mixture of features that early evolutionists seized upon in support of their new theory, because it clearly bridged the gap between warm-blooded, flying birds and cold-blooded, creeping reptiles. Reptile-like features include narrow jaws lined with tiny teeth (all modern birds lack teeth), clawed fingers on the wings (claws are absent from the wings of all living birds except the leaf-eating Hoatzin of South America) and a long, elegant bony tail (bird tails contain only a few stumpy vertebrae). By contrast, bird-like characters include the presence of a small, boomerang-shaped wishbone in the chest (a feature then unknown in any reptile, though now known in many non-bird dinosaurs) and, most importantly, feathers. Until the late 1990s, feathers were considered to be unique to birds, and most probably associated with the evolution of flight, so this was key to interpreting *Archaeopteryx* as the earliest bird. The plumage is obvious and beautiful, with impressions of long flight feathers surrounding the arms and a sheath of smaller feathers adding length and width to the tail. As more fossils have been recovered, many of the features once thought to be unique to birds have been found in other, earlier theropod dinosaurs, showing that the specialized body plans we see in modern birds did not appear all at once, but were assembled bit-by-bit over extended evolutionary timescales. As the gap between *Archaeopteryx* and other dinosaurs has shrunk, some palaeontologists have suggested it was not a bird at all, just a very bird-like dinosaur, although others still consider it the first true bird. One contentious area relates to its flight abilities, with some studies suggesting it was a glider (at best) and others that it might have been capable of short, rapid bursts of flight, although all agree that *Archaeopteryx* lacked the aerial grace of any modern bird. Nevertheless, despite these controversies, this tiny Jurassic dinosaur/bird remains incredibly important to our understanding of the origins of both birds and flight.

OPPOSITE: The unique combination of reptile- and bird-like features of *Archaeopteryx* gave Charles Darwin a stunning example of how evolution could transform one type of animal into another.

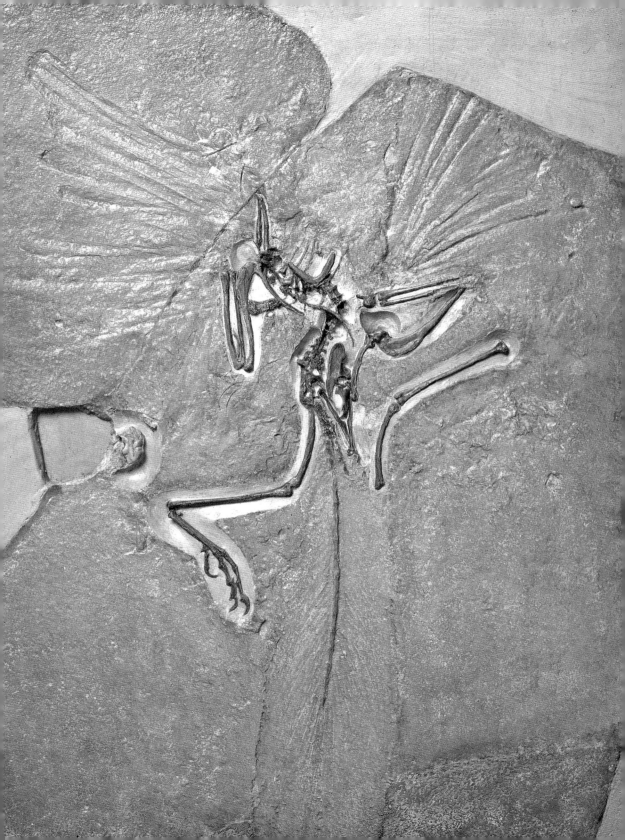

# Early birds diversify
## *Confuciusornis*

FOR MILLIONS OF YEARS, the sky had been the domain of pterosaurs and insects. However, a burst of evolutionary innovation during the Cretaceous Period saw increasing numbers of birds taking to the air. Until recently, Cretaceous bird fossils were rare, fragmentary and known mainly from close to the end of the period. However, new discoveries from Spain, Argentina and, especially, China are showing that Mesozoic birds were much more varied than previously realized. In particular, the spectacular ancient lake deposits of the Jehol Biota, in northwest China, dating from 131–120 million years ago, have yielded thousands of complete bird fossils, many preserving exquisite details like feathers and gut contents. Perhaps the most abundant of these is the magpie-sized *Confuciusornis*. Hundreds of complete skeletons have been recovered from the Jehol Biota, giving us unprecedented insights into its biology. Unlike its earlier cousin *Archaeopteryx*, but in common with modern birds, *Confuciusornis* lacked teeth and was one of the first birds to possess a horny beak. One specimen has a group of small fish bones in its throat, showing that fish formed at least part of its diet, and its strong beak is consistent with feeding on a wide range of foodstuffs. The wings still bear claws, but it was clearly a more adept flier than *Archaeopteryx*: its wings are longer, with large, well-developed, asymmetrical flight feathers for producing lift, and its breastbone is expanded to provide a greater area for the attachment of flight muscles. Overall, the wings are similar in shape to those of modern woodland-dwelling birds, suggesting that *Confuciusornis* was capable of complex manoeuvres. Elongate feathers took the place of a long, bony tail and the tail skeleton is reduced to a stump, like the 'parson's nose' of a chicken or turkey. The claws on the wings and on its feet indicate that it was probably capable of climbing and perching in trees. *Confuciusornis* provides us with one of the earliest examples of a phenomenon called 'sexual dimorphism' – the presence of clear differences in appearance between the sexes. Some *Confuciusornis* fossils have a pair of extremely long, ribbon-like tail feathers, but these structures are absent in others. As the ratio of long-feathered to short-feathered *Confuciusornis* specimens is about 50/50, it seems likely that we are looking at a population of males and females. It is assumed that, as in living birds, the 'showier' tail feathers were sported by males and might have been important in display. Living alongside *Confuciusornis* were a variety of other early birds ranging widely in size (from wren-sized to gull-sized), diet (fish eaters, insect eaters, seed eaters) and habits. These finds transformed our knowledge of early bird evolution, fine-tuning our understanding of modern bird origins.

This almost complete skeleton of *Confuciusornis* possesses beautiful feathers that show the outlines of its wings. The lack of long tail feathers suggests that this individual might have been female.

# Scaly skins
## *Haestasaurus*

**SKELETONS GIVE US VALUABLE** anatomical information, particularly with respect to growth, feeding, locomotion and evolutionary relationships, but there are many aspects of dinosaur biology and behaviour that cannot be gleaned from bones alone. Luckily, different kinds of fossils offer additional insights that help palaeontologists to reconstruct dinosaurs as living, breathing animals.

For example, fossil skin samples (or, more often, skin impressions) deliver clear, direct evidence on the outward appearance of a dinosaur and provide clues on skin function during life. As dinosaurs are reptiles, it was always expected that they would

be scaly, like living crocodiles, lizards and turtles. This was confirmed in 1852, when Gideon Mantell, an English country doctor, discovered small patches of dinosaur skin. While working on the arm bones of a new sauropod dinosaur from the south coast of England (later named *Haestasaurus*), Mantell noticed that the rock surrounding the animal's elbow was ornamented with a mosaic-like pattern, made up of many individual tile-like polygons that varied in size and shape. These structures were a perfect match for the patchwork of scales seen in modern reptiles.

Following Mantell's discovery, all dinosaurs were assumed to be scaled, and as more specimens were recovered from all over the world this idea was upheld again and again. Sometimes only small segments of skin were recovered, but the fossil fields of the USA and Canada soon yielded spectacular specimens of dinosaur 'mummies', usually duck-billed dinosaurs like *Edmontosaurus*, that were encapsulated in skin. These were not 'mummies' in the sense of Egyptian pharaohs (bandages were not available!) but were formed when the carcasses of these animals partially rotted, losing their soft internal organs but retaining their tough hide, which then 'shrink-wrapped' the skeletons. It is thought that such carcasses were exposed for long enough to dry out, hardening the skin and favouring its preservation after burial. These more complete examples show that scale size and shape can vary considerably over the body surface. Scales are often smaller and more densely packed near joints (presumably allowing greater flexibility), whereas on other parts of the body they can be grouped into intricate, rather beautiful patterns. These include rosette-like features that are composed of a large central scale surrounded by smaller ones. It is possible that different scales and/or scale patterns were differently coloured and these might have been important in display or camouflage. Some duck-billed dinosaurs had a line of low, ridge-like scales extending along the middle of the back and down along the tail, and the little ornithischian *Kulindadromeus* sported a variety of unusual scales, including ones with comb-like projections.

This is the first example of dinosaur skin to have been found anywhere in the world, from the Early Cretaceous sauropod *Haestasaurus*. The scales vary greatly in size and shape, even in this relatively small part of the animal's hide.

# Feathers and fuzz
## *Sinornithosaurus*

SPECTACULAR DISCOVERIES FROM the Lower Cretaceous Jehol Biota of China have revolutionized our understanding of bird origins. The unusual geological conditions that prevailed during the formation of these fossils involved frequent volcanic eruptions, whose ash and gases killed and quickly entombed the animals and plants living in, near and above the shallow lakes that once covered the region. This allowed the exceptional preservation of soft tissues, which are usually lost during fossilization, notably skin.

The first feathered dinosaur from China was unveiled in 1996 with the naming of *Sinosauropteryx*. Since then, around 50 non-bird dinosaur species have been found with confirmed evidence of a feathery covering. The wide variety of different skin structures preserved in these fossils has allowed scientists to reconstruct how feathers increased in complexity, and has provided insights into the reasons why they should have evolved in the first place. Many of these distinct feather types can be seen in a beautifully preserved specimen of a *Velociraptor*-like Jehol dinosaur, *Sinornithosaurus*, nicknamed 'fuzzy raptor', for obvious reasons.

A small part of the coat of 'fuzzy raptor' is made up of simple, straight, hair-like filaments. However, most of the body, including the neck, back, chest, thighs and tail, is covered with more complex structures, which are composed of several such filaments that arise from a single, combined base. This is akin to the basic branching pattern that we see in the downy feathers covering the bodies of living birds, and whose fluffy structure makes them so good at trapping heat (which is useful for staying warm and brooding eggs). However, the arms of *Sinornithosaurus* sport yet another feather type, more like the wing feathers of birds. These arm feathers have a central, hollow tube, called the shaft (or rachis), which supports many finer, shorter branches, called barbs. Together the barbs line up to form a continuous, flat surface on either side of the shaft – these surfaces are called vanes. The vanes in *Sinornithosaurus* are approximately symmetrical around the shaft. At first sight, these arm feathers look just like those on a bird wing, although there are subtle, but important, differences. First, in modern bird feathers the barbs are linked together by tiny, hook-like structures called barbules, which help to keep the

OPPOSITE: 'Fuzzy raptor' is one of the most important feathered dinosaur fossils. This small skeleton shows that a single dinosaur species could have diverse feather types, with distinctively shaped feathers in different body regions.

feather vanes rigid – these are absent in *Sinornithosaurus*. Second, flight feathers have one vane that is larger than the other and this asymmetry helps the bird create lift as it flaps its wings. This is because the shape of a bird feather causes air to move over the top of the wing faster than the air beneath it, and these differences in airflow help to 'pull' the bird up. So, although it had quite bird-like feathers, *Sinornithosaurus* was unable to fly, because it lacked this lift-generating capability, and other features of its anatomy – such as the lack of really large chest muscles, the lack of a special pulley system around the shoulder that helps wth the flight stroke, and that its arms were too short to provide a sufficiently large lift surface for its body mass – were also unsuitable for flight.

When we compare the varied feather types present in feathered dinosaurs (including birds) and combine this with information on their relationships, a clear evolutionary

OPPOSITE: This life reconstruction of *Sinornithosaurus* shows how the longer feathers were arranged along its limbs and tail, while the rest of the body was covered in shorter 'fuzz'. *Sinornithosaurus* might have been combining insulation from fuzz with the ability to use longer feathers for display.

sequence emerges. The earliest feathered dinosaurs (e.g. *Sinosauropteryx*) have simple filament-like structures. In more advanced theropods, these structures branch to form more complex, down-like structures. Adding more barbs to these produces feathers like those on the arms of *Sinornithosaurus*. Finally, adding barbules and changing the proportions of the vanes occurs in birds. These changes occur in the same order as those seen by biologists during the development of feathers in living bird embryos.

Although old ideas about feather evolution linked the appearance of feathers with the origin of flight, these new discoveries toppled this hypothesis. The earliest filament-like feathers clearly could not provide lift: indeed, feathers useful for flight did not appear until close to the origin of birds, and feathers are now known to have evolved long before the other modifications of the skeleton that are essential for flight. So, what were they for? Honestly, we are not sure, but there are two equally good ideas. It is possible that feathers evolved for insulation. Most feathered dinosaurs were small, and small animals lose heat quickly. A fluffy covering would have helped them to stay warm, meaning that they could remain active for longer, especially at cooler times of day. Moreover, parts of the Mesozoic world were chilly (at least at some times of year). For the same reason, feathers might have been helpful in brooding eggs. Alternatively, feathers might have appeared for display – it is notable that they are often prominently positioned on dinosaur arms and tails, parts of the body frequently used by living animals when showing off during courtship and territorial behaviours. Moreover, it is likely that they were coloured. A combination of these factors might have been at play, with advantages gained from insulation being amplified by an additional use in display, or vice versa.

As the Jehol Biota includes examples of many different theropod sub-groups, scientists are confident that all coelurosaurian theropods had feathers, and it is plausible that they were present in all theropods. Three ornithischians (*Tianyulong*, *Kulindadromeus* and *Psittacosaurus*) also have feather-like structures, as do some pterosaurs. However, feathers have not yet been found in any sauropodomorph or any dinosaur ancestor, and most ornithischians were clearly scaly. As a result, it is unclear if feathers were present in all dinosaurs or only some, and this remains an open question. To solve it, we need older deposits, from the Triassic and Jurassic, with the same exceptional skin preservation as these amazing Chinese fossils, which could prove whether or not earlier dinosaurs were feathered.

# Colour
## *Anchiornis*

PALAEONTOLOGISTS ILLUSTRATE their lectures with beautifully rendered reconstructions of dinosaurs, which are created by a growing band of dedicated palaeoartists. Given the realism of these artworks, I'm often asked "how do you know what colours they were?" Fifteen years ago, I had a stock answer: "sadly, we'll never know, as colours don't preserve in fossils, so we're left with our imagination." How wrong I was. Until recently, all known dinosaur fossils lacked evidence of colour. However, a few exceptional specimens have now broken this barrier. The first breakthrough came in 2010, from a little bird-like dinosaur from the Late Jurassic of China, *Anchiornis*. Using high-powered scanning electron microscopes, Chinese and American scientists examined its feathers in exquisite detail and found the impressions of small, rounded features on their surfaces. These were the remains of melanosomes – tiny, microscopic structures found throughout the bodies of all animals that contain pigment. Luckily for palaeontologists, the shapes and sizes of melanosomes in living animals are known to be related to the type of pigments that they contain, and this, in combination with the number of melanosomes, helps control skin, hair, scale or feather colour. This knowledge can be used to distinguish various darker colours: shades of brown, grey, black and red. By looking at the sizes and shapes of the melanosomes preserved in feathers from different parts of the body, the team were able to produce the first accurate colour reconstruction of an extinct dinosaur: in this case with greyish feathers covering much of the body and neck, long black feathers on the arms and, rather daintily, a crown of short, russet feathers. Following this stunning revelation, and using similar methods, a few other dinosaur fossils have offered up their secrets. *Archaeopteryx*, *Beipiaosaurus* and *Caudipteryx* were black or dark brown; *Sinornithosaurus* and *Sinosauropteryx* had colour banding including russet or ginger tones. Taking this work further, by looking at the fine-grained physical structure of the feathers themselves, it appears that some dinosaurs, such as *Caihong*, *Microraptor* and *Wulong*, might have glimmered with shiny, iridescent hues. Although most of our results so far have come from feathered theropods, the early horned dinosaur *Psittacosaurus* and the ankylosaur *Borealopelta* seem to have been dappled and 'counter-shaded' (dark on top, lighter underneath) for camouflage. So far, we have not been able to uncover the full range of colours seen in birds, including the most striking yellows, oranges, greens and blues, which have different physical and chemical foundations from the other known colours. However, who knows what might be discovered tomorrow?

These *Anchiornis* skeletons (above) were discovered in ancient lake sediments, explaining the presence of fish fossils alongside them.

It's tempting to speculate that the russet head crest and patterned feathers of *Anchiornis* might have been used for display.

# Soft tissues
## *Scipionyx*

STUDIES ON FOSSILIZED SKIN have overturned earlier ideas on dinosaur biology, providing critical information that could not be obtained from bones alone. Skin is just one of the 'soft tissues' present in any living animal; other examples include the brain, heart, lungs, digestive system and muscles. Sadly, unlike bones and teeth, which are made of tough, resistant minerals, soft tissues usually disappear quickly after death, as they are easily eaten by scavengers and predators or rot away through the actions of bacteria and fungi. As a result, palaeontologists usually have no choice but to speculate on their shape, size and function, using comparisons with living relatives – birds and crocodiles – to constrain their ideas, as well as the 'ghosts' that these structures leave behind on the skeleton, such as the tell-tale marks of muscle attachments on limb bones. Nevertheless, a handful of truly astounding fossils fill some of these gaps.

The most extraordinary of these is the only known specimen of a tiny theropod dinosaur, *Scipionyx*, which comes from the Early Cretaceous of Italy. Its partially complete, flattened skeleton, only 24 cm (9½ inches) in length, is splayed out on a slab of grey marine limestone revealing fabulous detail. However, even more remarkably, *Scipionyx* preserves a variety of soft tissues, which can be seen all over its body. These include bundles of muscle around the neck and in the area between the hips and tail, including parts of the major muscle responsible for providing power during walking. The individual muscle fibres are often visible, giving these areas a characteristic 'meaty' appearance. In the small gaps between the neck vertebrae, remains of the ligaments that would have connected them can still be seen, and a capping of cartilage covers the ends of many limb bones. Also made of cartilage, the circular rings that surrounded and supported the windpipe are stacked along the neck. When we examine the claws on the hands, it is clear that remains of their horny outer sheaths are present, which extend the true lengths of the claws well beyond their inner bony cores. Perhaps most impressively, this little dinosaur gives us our only example of a well-preserved dinosaur gut. A set of connected, tube-like structures loop their way through the animal's belly and it is possible to see their internal divisions, into the different parts of the small and large intestine, as well as the delicate folds of the intestinal walls.

Oddly, despite this wealth of detail, no skin is preserved, so we do not know if *Scipionyx* was scaled or feathered. Still, this amazing specimen is the closest we have come to the carcass of a living, breathing Cretaceous dinosaur, and it offers the hope that other similar fossils might yet come to light.

The abdomen of *Scipionyx* is filled with brown mineral deposits that trace the outline of its liver and intestines. Other coloured patches around the skeleton are remnants of muscular tissue.

# Dinosaur reproduction
## Titanosaur egg

DINOSAUR EGGS COME IN a spectacular range of shapes and sizes. Remembering that birds are part of the dinosaur radiation, the smallest dinosaur egg we know about belongs to a living animal, the bee hummingbird. Each of its tiny eggs is roughly the size of a pea and weighs only 0.05 grammes (0.0018 oz). Perhaps surprisingly, eggs at the other end of the scale did not belong to gigantic sauropods, but to a more recently extinct species, the elephant bird of Madagascar (*Aepyornis*), whose eggs were up to 40 cm (15¾ inches) long and had a volume of nearly 13 litres (27½ pints) (equivalent to over 10 kg (22 lb), roughly seven times the size of an ostrich egg). Although the eggs of some extinct dinosaurs, such as those of giant oviraptorosaurs were longer than those of elephant birds, they were much narrower in shape, lacking the typically oval shape of bird eggs: as a result, they had much smaller volumes. One such egg type is called *Macroelongatoolithus* (the name meaning, literally, 'large, elongate, fossil egg'), which reaches an impressive 60 cm (24 inches) in length.

Sauropod eggs were average by comparison, reaching up to 20 cm (8 inches) across, the size of a cantaloupe melon. This titanosaur egg from the Late Cretaceous of India is a typical example, although this particular specimen has an interesting history. It was discovered sometime in the 1840s (exact date unknown) and donated to the Natural History Museum, in London. However, dinosaurs were only established as a group in the 1840s and no one was yet thinking about dinosaur eggs. As this specimen is filled with beautiful purple and white layers of the mineral agate, which grew in the egg as it became fossilized, the original owner assumed that it was just an interesting rock. Consequently, it was never seen by the museum's palaeontologists, but was added to its Mineralogy Collection. It was only in 2021 that curators noticed some unusual features that quickly led to it being recognized as a dinosaur egg. Not only was it a dinosaur egg, but its date of collection suggests that it might have been the first example of a dinosaur egg to end up in a museum!

The features that identified this object as an egg are found in all other fossilized dinosaur eggs. Luckily, the egg is preserved in two halves so we can see its internal structure. The outer part of the specimen is formed by a narrow layer of even thickness

OPPOSITE: The surface of this titanosaur egg is ornamented with thousands of tiny 'pimples' that give the shell surface a roughened texture.

Although the interior of this egg is now filled with brightly coloured agate, it would have originally contained a yolk, egg white and a tiny dinosaur embryo.

(about 4.5 mm (c.1/6 inch)), which surrounds the whole structure. The mineral-filled interior is spherical in shape, matching the overall shape, and was originally hollow. The outer layer has tiny holes (called pores) and its surface is ornamented with many small, low, rounded bumps. Pores are essential for letting life-giving oxygen into the developing egg and letting the waste carbon dioxide out, and the complex surface patterns might have assisted in keeping the pores free from dirt. The spherical shape, surface patterning, and even the thickness of the outer layer are features that are consistent with features seen in fossil eggs that were found in the same area, more recently. These Late Cretaceous fossil beds in India are rich in the remains of titanosaurs, suggesting that they were the likeliest egg-layers.

This brings us to a problem. Fossil eggs are not usually found inside their mother's bodies, but in nests (although there are two or three very rare exceptions), and almost all eggs and nests have, so far, been found in sites lacking dinosaur bones. So, how do we know which egg goes with which dinosaur? Sometimes, we are lucky and the eggs contain the fossilized remains of an unhatched baby dinosaur. In these cases, we can often identify the exact species involved or at least the dinosaur group to which the egg belonged. However, such finds are scarce (our titanosaur egg does not contain any bones, sadly). Luckily, there are other clues. It turns out that when we list features like egg size and shape, and more subtle features such as the microscopic structure of the eggshell, for those eggs that do contain hatchlings, some useful patterns emerge. These comparisons show that each major dinosaur group laid slightly different egg types, each with their own unique set of characteristics. For example, most theropod eggs are almost sausage-shaped – long and narrow, with rounded ends. By comparison, most sauropod and hadrosaur eggs are round. Microscopic structure, in particular the shapes and orientations of the individual calcite crystals of which the shell is made, is particularly useful for palaeontologists, because you do not even need a whole egg, just a fragment, to identify the type of dinosaur that you are dealing with.

Detailed study of eggs can reveal other secrets of dinosaur biology. For example, living crocodiles, turtles and lizards lay plain, white eggs, whereas birds produce a variety of shades ranging from beige to vivid blues and greens, as well as eggs decorated with many different patterns. This ability assists in camouflage, reducing the risk of egg predation, and helps the parents to recognize their own eggs. Recently, a range of advanced chemical analyses were applied to dinosaur eggs, revealing their original colours. Some, like those of hadrosaurs and sauropods, were white, similar to their reptilian relatives. However, theropod eggs, like those of *Deinonychus*, share the same blue-green colours seen in birds, and were probably dappled. This is yet another example of a feature formerly thought to be unique to birds that originated much earlier in their dinosaur ancestors.

# Parenting
## *Maiasaura*

DINOSAUR NESTS WERE first uncovered in the Gobi of Mongolia during the 1920s, confirming that they laid eggs. Although these discoveries caused a sensation worldwide, they did not immediately lead to any surprising revelations. After all, dinosaurs were known to be reptiles and most living reptiles lay eggs and some build nests. Dinosaur eggs (and a few nests) continued to be found in the years that followed, from many different sites, but it was not until the discovery of 'Egg Mountain' in 1977, in the remote badlands of Montana, USA, that nests would give major new insights into dinosaur behaviour.

Excavations at Egg Mountain were to provide the first compelling evidence of a complex social life for any dinosaur species. Hundreds of bones belonging to a single species of hadrosaur were unearthed, ranging from tiny hatchlings, which would have easily curled up in the palms of your hands, to adults that were 9 m (29½ ft) in length. The bones were found alongside multiple nests, each containing a mixture of hatched and unhatched eggs. The nests are shallow, bowl-like structures, approximately 2 m (6½ ft) across, and would originally have been made of mud (now compacted into rock), with a raised outer rim surrounding a central hollow housing the eggs. There are around 15–20 eggs per nest and each egg is 10–12 cm (4–5 inches) in diameter. Unlike earlier discoveries of nests, which were found singly, there are many nests at Egg Mountain, which are spaced around 7 m (23 ft) apart, which is slightly shorter than the length of one of the adult hadrosaurs found nearby. Moreover, the nests were often stacked on top of each other, separated by layers of mud, with newer nests overlying older, abandoned nests.

Taken together, these observations suggest that Egg Mountain was once a communal nesting site for this hadrosaur, a place where adults would come together to lay their eggs. The tight packing of nests was probably for protection – having lots of large hadrosaurs in one spot would allow them to spot predators coming and potentially put up some kind of defence – and having lots of nests together lessens the chances of *your* nest being the one that a predator might select. The vertical stacking of nests suggests that the hadrosaurs returned to this spot year after year: maybe there was something about this particular spot that also helped to deter predators. Even more intriguing is the mixture of baby and adult dinosaurs. When examined in detail, the leg bones of the babies are not fully formed, suggesting that they were not capable of leaving the nest.

A reconstructed nest of *Maiasaura* hatchlings based on the discoveries made at Egg Mountain, Montana, USA.

However, their tiny teeth show signs of wear, proving that the hatchlings were feeding, but the nesting sites had no obvious source of fresh plant food. The helplessness of the hatchlings and the presence of adults lead to one conclusion: the adults must have been bringing food to the nests. At the time these discoveries were made, such advanced parental care was considered unique to birds and mammals, so the idea that dinosaurs might have shown similar behaviours was revolutionary. It also gave this new species of hadrosaur its very fitting name: *Maiasaura*, the 'good mother lizard'. The recognition of these complex behaviours was one of the key factors that helped to persuade palaeontologists that dinosaurs were dynamic, sociable animals.

Since then, other dinosaur nests have been found that have given further insights into dinosaur parenting. *Maiasaura* was too heavy to sit on its own eggs, and rather than brooding them it would have covered the eggs with vegetation to help keep them warm. However, some smaller dinosaurs definitely did sit on their nests. The first found dinosaur nests in the Gobi were originally thought to belong to the early horned dinosaur *Protoceratops*, mainly because this is the commonest dinosaur at those sites. However, further work on the nests showed that this was not the case. Some '*Protoceratops*' nests have skeletons of small theropods associated with them. Initially, these theropods were

assumed to be egg predators and were given the name oviraptorosaurs ('egg-stealing lizards') as a result. But if we look again, it turns out that these small theropods, including *Oviraptor*, *Citipati* and *Nemegtomaia*, were not attacking the nests – they were sitting in the middle of them, their arms curled around the eggs. When palaeontologists looked into the eggs they found baby oviraptorosaurs inside, not *Protoceratops*. Rather than getting a quick meal, it turns out that these species were caring for the eggs, brooding them, using the same kind of technique as a living bird. It is a shame that a group of theropods that were clearly doting parents will forever be labelled with a name that suggests the opposite.

Although there is strong evidence that some dinosaurs provided ample care for their nests and young, other dinosaurs were less nurturing. In particular, sauropods were not devoted parents. Sauropod nests are mere scrapes in the ground, and there is no indication that the adults spent time with the hatchlings. It should be borne in mind, however, that an adult sauropod is many thousands of times larger than its hatchlings. There is no way that a sauropod could brood its eggs without breaking them and a huge adult might easily have stepped on a tiny hatchling by accident. In this case, the best thing for a parent to do would be to lay lots of eggs and hope that at least a few of the young survived independently.

Thanks to the discoveries of nests and eggs, palaeontologists have been able to deduce how quickly *Maiasaura* grew up and how it's body changed as it got older.

# Sexing a dinosaur
## Oviraptorosaur

MUSEUMS OFTEN GIVE THEIR superstar dinosaur specimens nicknames, endowing them with personality and helping visitors to relate to them. Perhaps the most famous example is the Field Museum's beautiful *Tyrannosaurus*, 'Sue'. Sue is named in honour of its discoverer, fossil hunter Sue Hendrickson, but many people assume that the skeleton is female too. However, to be honest, we do not know if 'Sue' was male or female – it is surprisingly difficult to sex a dinosaur.

This is for two main reasons. First, as soft tissues usually disappear during fossilization we have never found the reproductive organs of any dinosaur, so the most obvious way to tell the difference is missing. Second, among living reptiles and birds, there are few features of the skeleton that distinguish males and females. As a result, palaeontologists lack clear guidelines that they can apply to fossil remains.

Today, male mammals, birds and reptiles are generally larger or bulkier than females. So, could we use body size or build to spot a male or female dinosaur? Sadly not. These 'rules' on size are easily broken – for example, female birds of prey and female ostriches are larger than the males. Also, fully grown adults are needed for fair comparisons between skeletons and we do not always know if a particular dinosaur specimen had stopped growing (after all, a small individual might either be female or a male that is not fully grown). In addition, ideally, you would need to demonstrate that around half of the adults known for any dinosaur species were clearly smaller or larger than the others, reflecting the 50/50 sex ratio of males and females that we see in living populations. Sadly, however, we usually have only a few specimens to go on, so the large sample sizes needed to prove this are lacking.

This exceptional fossil of an unnamed oviraptorosaur from China contains at least three unhatched eggs (the light grey oval objects seen in the centre and to one side). The condition and position of the eggs shows they weren't dinner but were ready to be laid, so we know this individual was female.

Turning again to living animals for help, adult male mammals and birds often have display structures that they use for showing off or fighting during mating, such as the antlers of deer or the tail of a peacock. Perhaps finding similar features in dinosaurs would help? Again, in most cases, sadly not. We run into the same issue of sample sizes that are too small to find clear differences, and in the few examples where there are lots of specimens to measure and compare (ceratopsian frills and horns, for example) there are no obvious differences in these features that could be linked to sex. So far, there is only one clear example of a sexual difference among Mesozoic dinosaurs: the presence of long tail feathers in what are assumed to be the males of the early bird *Confuciusornis*.

Detailed bone structure has offered some potential clues on sex. Bones are composed of many different tissue types and female birds make a special type of bone, called medullary bone, which forms inside their hollow limb bones. Medullary bone is used as a temporary store for calcium, which the bird breaks down for use in eggshell formation during egg-laying. Medullary bone has a characteristic spongy appearance under a microscope, which is easy to distinguish from other bone types. By looking at thin sections of dinosaur bone, palaeontologists have used these features to identify several specimens where medullary bone is present, which suggests that these individuals might have been females that were about to lay eggs. However, it must be remembered that medullary bone is a transient feature that is only present in adult females some of the time: it is absent in females that are not laying eggs. As a result, it can only be used to identify egg-laying females and cannot distinguish non-egg-laying females from males or juveniles. Moreover, bone tissues similar to medullary bone have now been found in other parts of the skeleton, in areas where they cannot have been used as a store for eggshell calcium, so even this promising method is not fool-proof.

A final method that has been tried looks at dinosaur tails. In male crocodiles, special muscles anchor their reproductive organs to the base of the tail and these are absent in females. Where present, these muscles extend back to the base of the tail; as a result, some of the tail bones in male crocs are larger than the corresponding bones in females. Palaeontologists have tried to find these differences in the dinosaur tail bones, but to no avail.

Currently, the only reliable way to sex a dinosaur depends on one very rare circumstance: finding a skeleton that contains eggs. To rule out the possibility that the eggs were food, which is the other plausible way they might end up inside a dinosaur, the eggs have to be complete (uncrushed by the animal's jaws or gut muscles), lack evidence of digestion, and lie within the skeleton in a way that is consistent with dinosaur reproductive anatomy, namely deep in the animal's posterior end. To date, only a handful of dinosaur specimens have been found that contain convincing evidence of unlaid eggs, including one specimen of *Sinosauropteryx* and an unnamed oviraptorosaur from China. However, although these fabulous fossils provide us with our first definite female dinosaurs, they offer no other clues on how to distinguish males and females using more obvious features of the skeleton.

Finally, it is worth remembering that some of the dinosaurs on display in museums are composites, made by combining less complete individuals of the same species together. In these cases, they probably include elements from both male and female skeletons. Sexing a dinosaur remains one of the biggest challenges for palaeontologists.

# Life at extremes
## *Patagotitan*

DIPLODOCUS. TRICERATOPS. TYRANNOSAURUS, some of the most famous dinosaurs and the ones that most often come to mind, all have something in common: enormous size. To many, this is the defining feature of dinosaurs and was one of the formal criteria used by Richard Owen, a British anatomist and palaeontologist, to distinguish them from other reptiles when he named the group in 1842. If we ignore birds for a moment, the average size of an extinct dinosaur is about 1 tonne (1 ton) – the size of a walrus – so large size was clearly an important part of their biology. However, this masks the fact that dinosaurs started their evolutionary history as relatively small animals, only a couple of metres in length, and that they went on to occupy one of the greatest ranges of body size achieved by any animal group. Body size is fundamental to biology: it influences diet, territory size, movement, ecology and reproductive rate, as well as many other aspects of behaviour and metabolism. The smallest dinosaur is one that is alive today – the bee hummingbird, from Cuba, which reaches a maximum weight of 2 g (0.07 oz), about the same as a paperclip. If we consider only extinct dinosaurs, there is a diversity of small species that would easily slip into a bag. These include theropods like *Anchiornis*, *Microraptor* and *Parvicursor* and the heterodontosaurid *Fruitadens*, which were 35–65 cm (1¼–2¼ ft) long, roughly the size of a magpie, and weighed 100 g (4 oz) to a kilogramme (2.2 lb). This is considerably less than the 1,000,000 g (2,020½ lb) that the 'average' dinosaur weighed. At the other end of the scale, we have massive sauropods, the largest animals ever to walk the Earth, which easily overshadowed the largest land mammals. Among these, supergiants like *Patagotitan*, which is the most completely known of these behemoths, probably weighed around 60 tonnes (66 tons) or 60,000,000 g, the same as 30,000,000 bee hummingbirds! More fragmentary remains, like those of *Argentinosaurus*, have led to even higher body mass estimates of up to 100 tonnes (110 tons), although these have recently been scaled down to 70 tonnes (77 tons), with much uncertainty due to the incompleteness of the material and disagreements around the calculations needed to compute dinosaur weights. The only other animals to occupy such an expansive size range are the mammals, ranging from the diminutive bumblebee bat (weighing 2 g (0.07 oz)) to the mighty blue whale (up to 150 tonnes (165 tons)). Whales are the only animals to outclass dinosaurs in terms of size, but have the advantage of living in the oceans, where there are abundant high-protein food sources (squid or krill) for fuel and where gravity places fewer demands on the strength of muscle and bone.

The largest titanosaurs, like *Patagotitan*, are known only from Argentina. Scientists still have no idea why so many giants should have evolved in this particular region.

# Growing up fast
## *Ngwevu*

HISTORICALLY, IT WAS ASSUMED that dinosaurs grew slowly, like today's reptiles, and that it must have taken them many decades to reach final adult size, implying lifespans of a century or more. However, there is now overwhelming evidence that dinosaurs benefitted from extremely rapid growth, more similar to that seen in mammals and birds, and that even the largest dinosaurs reached their maximum size in around 25–30 years.

Our insights into dinosaur growth come from a rather surprising source: microscopic bone anatomy, a field called osteohistology. Osteohistologists cut wafer-thin sections from bones, only a few hundredths of a millimetre thick, to look at the different types of bone tissues and structures inside and how these are distributed through the bone. By using high-powered microscopes that shine light through these almost transparent slices, they can describe features like the tiny cavities left behind by the bone-forming cells, minute blood vessels and the fine fibres that anchor muscles to bones. More importantly, bone internal structure also preserves an accurate record of growth. Each year, a new layer of bone tissue is added as an individual bone gets larger, and these layers reflect periods of faster and slower growth, which are controlled by factors like seasonal changes in food availability, temperature and weather. Each annual cycle is separated from the next by a period when growth effectively stops, and this temporary break in growth leads to the formation of a distinctive ring-like structure in the bone. By counting these rings, known technically as 'lines of arrested growth' (or LAGs), osteohistologists can determine how old an animal was when it died, in a way analogous to aging a tree using the rings in its trunk. In addition, the distance between a pair of LAGs is related to the amount of growth that took place, so measurements of this distance allow the speed of growth to be calculated. Put simply, a thick bone layer between two LAGs suggests high, rapid growth, whereas a thin layer suggests slow growth. Thanks to work on living animals, the type of bone tissue deposited between the LAGs can also give important clues on growth rate. In slow-growing animals, the bone tissues are stacked in neat, regularly arranged layers; conversely, in fast-growing animals the bone has a chaotic, woven appearance. Luckily for palaeontologists, these details are well preserved in fossil bone, so we can use the same techniques to reconstruct growth patterns in long-extinct animals. Limb bones, especially thigh bones (the femur), shin bones (the tibia) and upper arm bones (the humerus), usually preserve the best record of growth due to their simple, tubular profiles, which maintain roughly the same shape as the animal gets bigger.

When osteohistologists trained their microscopes on dinosaur bones, they found several surprises. First, almost all the dinosaur specimens that have been examined in this way were still growing at the time of death. This suggests that although many of these individuals had reached impressive sizes, and must have been adults, they were still capable of further growth. Oddly, it seems that dinosaurs often died before they reached their full growth potential. Nevertheless, these still-growing adults must have been capable of reproducing, proving that dinosaurs did not need to reach a 'final' size to become parents, contrasting with birds, which only reproduce when growth has all but ceased.

A thin section of bone from the leg of the early sauropodomorph *Ngwevu*. Note the growth rings in the outermost layer of the bone.

Second, none of the dinosaurs sampled so far are more than 35 years old. The old idea of centenarian dinosaurs is clearly wrong. In general, larger dinosaurs reached greater ages, with sauropods and *Tyrannosaurus* known to reach 30+ years in age, examples of medium-sized hadrosaurs in their 20s, and small dinosaurs, like *Lesothosaurus*, with maximum documented ages of 5–6 years. It should be remembered, however, that we have not studied every single dinosaur bone in this way, and there might be older

individuals that osteohistologists have yet to study. It seems reasonable to speculate that the largest dinosaurs lived into their 50s, or maybe longer, comparable to many large animals today.

Third, dinosaur growth was super-fast. For example, an *Apatosaurus* was able to reach a body mass of 25 tonnes (27½ tons) in about 15 years, putting on a maximum growth spurt of nearly 5 tonnes (5½ tons) per year. Dinosaurs did not sustain the same growth rate throughout their lives but, like humans, had a 'teenage' growth spurt, when they increased in size most rapidly, with periods of slower growth straight after hatching and again towards adulthood. Indeed, many of the dinosaurs sampled so far are juveniles, whose bone microanatomy shows that they were still in this rapid growth phase.

Finally, detailed studies on multiple individuals from the same dinosaur species show that their growth could be highly variable, and that there were no set 'rules' on final adult size. For example, in the early sauropodomorph *Massospondylus*, some individuals that had nearly stopped growing are smaller than others that were still growing rapidly when they died. We are familiar with this phenomenon too – adult humans come in all shapes and sizes.

Taken together, all of this means that dinosaur populations were not made up of old, fully grown adults, but consisted largely of young adults and juveniles. This implies that dinosaurs had high mortality rates, which were probably offset by breeding early and by high reproductive rates. In many ways they were the rock stars of the Mesozoic world, living fast and dying young.

It used to be thought that *Ngwevu* was a juvenile specimen of the closely related *Massospondylus*, but careful comparisons between its growth record and skull shape showed that it was a distinct dinosaur species.

# Ecto- or endothermic?
## *Diluvicursor*

DINOSAUR METABOLISM HAS been the subject of constant speculation. For much of the nineteenth and twentieth centuries they were envisioned as being 'cold-blooded', like today's reptiles; that is, lacking sources of internally generated, metabolic heat, relying instead on external sources of warmth to raise body temperature (a condition termed 'ectothermy' – meaning 'outside heat'). During the 1970s and 1980s an alternative perspective was put forward as new aspects of dinosaur biology were revealed. Many of these, such as the potential for complex behaviour, rapid growth and high athletic prowess, which were exemplified by active-looking animals like *Deinonychus*, led some palaeontologists to argue that dinosaurs were 'warm-blooded' instead, like birds and mammals, maintaining a constant internal body temperature thanks to elevated metabolic rates (the 'endothermic' condition – 'inside heat'). The ever-closer links between dinosaurs and birds also supported this notion.

There is no definitive answer to this question, because we are not able to pop a thermometer under a dinosaur's tongue, but most palaeontologists agree that dinosaur physiology differed from that of living reptiles and that they must have had higher metabolic rates. Perhaps the most persuasive evidence comes from their rapid growth: similarly fast growth is seen in birds and mammals, related to their high metabolism, whereas slow growth characterizes reptiles. Another more recent line of evidence comes from the discoveries of feathered theropods, as it has been suggested that feathers might have evolved to help these small dinosaurs to trap internally generated heat to maintain high constant body temperatures.

Fieldwork in Alaska, Australia and Antarctica has revealed a diverse array of dinosaurs in the polar regions that lived there all-year round, and herds of hadrosaurs and ceratopsians from the Arctic show that they could be present in large numbers. Although these areas were not buried under ice during the Mesozoic, they were close to their current northerly and southerly positions. Geological evidence shows that these regions would have experienced severe winters, with temperatures close to freezing for months at a time. Only endothermic animals can survive in these conditions today.

It is also possible that the largest dinosaurs maintained a constant high temperature 'by accident'. The larger an animal is, the harder it is for it to lose heat from its body surface. As a result, any heat generated by muscular activity (e.g. walking) or digestion might have been trapped internally to raise the dinosaur's temperature.

This phenomenon is termed 'gigantothermy'. Finally, and most recently, sophisticated analyses of dinosaur eggshells have found further evidence for dinosaur endothermy. The exact chemical composition of eggshell reflects the body temperature at which it formed, so can be used as a kind of 'palaeothermometer'. This method suggests that the eggshells formed in warm-blooded bodies.

The small ornithopod *Diluvicursor* comes from southern Australia, which was situated well within the Antarctic Circle during the Early Cretaceous. It would have had to endure low temperatures for a least part of the year.

# Take a deep breath
## *Ornithopsis*

AS PALAEONTOLOGISTS OFTEN lack direct evidence for the shapes and sizes of dinosaur hearts, guts and other organs, they are forced to assume that they were similar to those of their closest living relatives, birds or crocodiles. This raises obvious problems because the organs present in these groups often differ. For example, the major blood vessels leaving the heart have quite different arrangements. Which should we choose for dinosaurs? Luckily, one major organ system leaves its mark directly on the skeleton, so we are able to reconstruct parts of it with high accuracy. This is the respiratory system, consisting of the windpipe, lungs and an extensive set of associated structures, called air sacs. Although the windpipe and lungs rarely leave clear traces, air sacs leave numerous clues on their size, shape and position, providing essential information for understanding dinosaur breathing.

Sauropod and theropod vertebrae, like those of *Ornithopsis*, have the same basic layout seen in all other backboned animals. A spool-shaped main body forms the lower part of the vertebra, which is attached to a plate-like upper part that bears knobs and processes for the attachment of back muscles and to link it with its neighbours. However, in both of these dinosaur groups (including birds) the main body of the vertebra usually bears large holes on its sides, which lead into honeycomb-like chambers within the bone. In addition, the upper part becomes more complex, bearing thin ridges and struts of bone, which are separated from each other by smoother, hollow areas. These holes, struts and hollows are not for decoration – they were once occupied by air sacs. In living birds, there are several sets of air sacs, situated in the neck, chest and trunk. They leave the same patterns on the bones nearby that we see in their dinosaur relatives.

Air sacs are balloon-like extensions from the lungs, which vastly increase the overall volume of the respiratory system. They are critical in storing and moving air, forming part of a one-way airflow system. This one-way system makes the uptake of oxygen and dumping of carbon dioxide incredibly efficient, by keeping air moving through it at all times (not just when breathing in or out) and by preventing mixing of oxygen-rich and used air during breathing. Such high efficiency gives birds the oxygen supply necessary for flight (and their hollow, air-sac-filled bones are very light, which also helps), allowed sauropods their long necks (which had lengthy windpipes that hold more air than the lungs alone) and permitted theropods and sauropods to reach large body sizes. It is not clear if all dinosaurs had this system however, because ornithischians and most early dinosaurs lack any bony traces of air sacs.

This vertebra, which formed part of the the trunk section of the backbone, is from the Early Cretaceous UK sauropod *Ornithopsis*. Note the large openings in the side of the vertebrae and the many hollows and struts. All of these openings would hosted air sacs when the animal was alive.

# Speed and gait
## Theropod trackway

SKELETONS PROVIDE HUGE amounts of useful information, but some aspects of dinosaur behaviour cannot be deduced from bones alone. Luckily, other types of fossils fill these gaps. The most abundant dinosaur fossils of all are not bones but footprints, which can occur as single tracks, as trackways composed of multiple prints (made by one individual), or at sites where many hundreds of prints are present forming numerous trackways, made by multiple dinosaur species that lived alongside each other. Although a single dinosaur had only one skeleton, it would have made many thousands, perhaps millions, of tracks during its lifetime, so these are a rich source of evidence.

The geological conditions needed to preserve tracks and bones are somewhat different, so they are rarely found together in the same rock layers. However, this means that tracks provide us with valuable insights that complement the bone record. For example, even if bones are totally absent, the presence of tracks can show that dinosaurs lived in a particular area at a particular period, providing useful information on their geographical and geological distribution. More importantly, tracks are formed by an interaction between the hand, foot or tail of a dinosaur and the sand or mud of the riverbank, lakeside or beach on which it walked. Each track captures an instant in time, providing a direct record of what a particular dinosaur was doing at that exact moment. For this reason, tracks give us some of our best information on dinosaur behaviour.

The most impressive dinosaur tracksite in the UK was found at Ardley Quarry, just outside Oxford, in the late 1990s. Quarrying for limestone exposed a series of Middle Jurassic rock layers that were covered with hundreds of footprints. These were arranged in 42 separate trackways that were made by sauropods (38 trackways) and theropods (four trackways), catching a moment when they walked along the shore of what was once a warm, shallow lagoon. Some of the trackways are up to 180 m (197 yd) long! We can identify these trackmakers based on a combination of features, as each dinosaur group has its own distinctive set of hand and foot shapes. Theropods were bipedal, so their trackways lack handprints, and their feet have three toes, all pointing forwards or slightly outwards, with the central toe longest. Each toe ends in a sharp claw, so the tip of each toe print is pointed and triangular. By contrast, sauropods walked on all fours so their trackways include both handprints and footprints. Sauropod fingers are bound together in a tube-like arrangement, so that when pressed into the ground their hands form a crescent-shaped or horseshoe-like print. Sauropod feet are differently built, with

These large, three-toed footprints form part of a trackway that was made by a *Megalosaurus* (or a close relative). They formed when the animal walked along a muddy shoreline in what's now Oxfordshire, UK during the Middle Jurassic period.

a large fleshy pad supporting the toes underneath, rather like that of an elephant. As a result, the footprints are large and circular in outline, and the front edge of the print includes small impressions from the blunt toe claws.

Other dinosaur groups possess their own distinctive combinations of hand and foot features that can be used to narrow down the identity of a trackmaker. However, it is usually not possible to match a footprint to a particular dinosaur species, for two reasons. First, the hands and feet of many different species within the same groups look pretty much the same. So, for example, we can usually tell if a track belongs to a theropod, but it might be impossible to narrow it down further than this. Second, with a single exception (one example of *Protoceratops* found in Mongolia), no dinosaur has been found dead at the end of a trackway, so there are no direct links between the track and the trackmaker.

Measuring the length of each individual track can tell us the size of the animal that made it, as there is a relationship between the length of the foot and the overall length of the animal's leg. For example, some of the Ardley theropod tracks are over 70 cm (2¼ ft) long, and this suggests that the trackmaker's hips were about 2 m (6½ ft) above ground level, which is consistent with a *Megalosaurus*-sized animal (8–9 m (26¼–29½ ft) in overall length). More interestingly, measuring the distance between two footprints in the same trackway can give an estimate of stride length. Thanks to work on living animals, which has shown how stride length is related to the speed of movement, taking many such measurements from a trackway allows us to calculate a dinosaur's speed. Using this method, one of the theropods at Ardley is thought to have reached a maximum speed of around 29 kmph (18 mph).

Trackways also illuminate other kinds of behaviour. At Ardley, the many sauropod trackways that cross the site are oriented in the same direction, suggesting that these animals were travelling together as a herd. Many other examples of herding have been discovered at tracksites around the world, especially for sauropods and hadrosaurs, suggesting that this behaviour was common among plant-eating dinosaurs. A few tracksites preserve evidence of more specific interactions. Some show that the smaller (presumably younger) herd members were positioned in the centre of the group, which was probably for defence. Other tracksites show examples of theropods following plant-eaters, providing a glimpse of their hunting behaviour. Yet more tracks have been interpreted as evidence for swimming, resting (shown by the dents made when squatting on the ground), and even mating, so they have the potential to unveil many aspects of dinosaur lives that would otherwise be unknowable.

A footprint from an ornithopod dinosaur from the Middle Jurassic of the Isle of Skye, Scotland.

# On all fours
## *Tuojiangosaurus*

DINOSAURS BEGAN THEIR evolutionary journey as bipeds, with the earliest members of each major lineage walking on two feet. Paradoxically, having gained this rare ability, several groups went on to abandon bipedality, reverting to the four-legged, quadrupedal stance seen in most other land vertebrates. This occurred on at least four independent occasions – in the ancestors of sauropods, hadrosaurs, armoured dinosaurs and ceratopsians – at different times.

Various changes were required to the forelimbs, changing their original grasping function to become supports for walking. Most importantly, the orientations of the shoulder, elbow and wrist joints had to change so that the hands could be rotated and placed palm-down to make contact with the ground. The arms became longer and closer in length to the legs, allowing easier coordination during walking and holding the head well clear of the ground. In most cases, the hand claws were modified into blunt hooves, better suited for spreading weight.

Sauropod quadrupedality is perhaps the easiest to explain. In this case, the shift to walking on all-fours coincided with major increases in body mass. Quadrupedality was probably essential for extra support, with the arms bearing a share of body weight. This idea is backed up by considering the sizes of bipedal dinosaurs: the largest bipeds, like *Tyrannosaurus*, weigh 7–8 tonnes (7¾–8¾ tons), whereas most sauropods are 15 tonnes (16½ tons) or more. You need to be a quadruped to become a giant.

It is much harder to explain the evolution of quadrupedality in the other three groups. Weight was probably a less significant factor, as armoured dinosaurs, ceratopsians and hadrosaurids became quadrupedal at lower body sizes, with the first animals to achieve quadrupedality in each group tipping the scales at no more than a few hundred kilogrammes. The additional burden of armour in ankylosaurs and stegosaurs has been suggested as a possible driver for becoming quadrupedal in this group, with the extra bony mass forcing them on to all fours. However, calculations of armour weight show that it was not a major contributor to overall body mass, so this idea seems unlikely. Intriguingly, quadrupedality in ceratopsians evolved at the same time as their huge skulls and frills, and it is possible that the mass of their enormous heads might have required them to use their arms to support the front of the body. Notably, all of these groups are plant-eaters, and it is tempting to speculate that advanced herbivory, requiring longer, more complex guts, might have been important in driving quadrupedality. Such lengthy guts could have affected the way in which weight was distributed within the trunks of these animals, perhaps requiring additional front-end support. Nevertheless, to be frank, we are still not sure why these three groups gave up bipedality, and palaeontologists are still seeking a clear answer to this question.

Armoured dinosaurs like this stegosaur – *Tuojiangosaurus* from the Late Jurassic of China – were one of four dinosaur groups to independently become fully quadrupedal.

# Predation
## *Proceratosaurus*

THE FIRST DINOSAURS WERE carnivores or omnivores, and a meat-based diet was retained by almost all theropods. It is also a common diet in living birds, from condors to shrikes. Meat has the advantages of being high in energy, as it consists largely of protein with a little fat, and being easy to digest, without the need for lots of chewing or long, convoluted digestive systems. Prey availability is the major challenge for carnivores, as no animal wants to end up as a predator's lunch. Considerable effort is often required to catch and subdue a potential meal, which is hard work that takes up most of a carnivore's time and energy. Fuelled by their meaty diets, theropods became the apex land predators of the Mesozoic. The type of prey taken depended on predator size: diminutive theropods like *Microraptor* and *Anchiornis* fed on insects, lizards and amphibians, whereas giants like *Carcharodontosaurus* and *Tyrannosaurus* ate other large dinosaurs, including sauropods, hadrosaurs and ceratopsians. The larger the predator, the more it needed to eat to survive, especially if there were long gaps between meals.

With only a few exceptions, most theropods had narrow, triangular teeth, whose tips were curved backward, with edges lined with dozens of small, sharp serrations. This basic design, a combination of hook and steak knife, was useful for both gripping prey and slicing through flesh efficiently. A handful of carnivores had more specialized teeth, such as the conical, spike-like teeth of spinosaurs, useful for impaling wiggling fish, and the thickened teeth of some tyrannosaurs, ideal for puncturing bone. Similarly, all theropods had large, strongly curved claws on their hands and feet, which would have been useful for gripping prey during hunting and in dismembering the unfortunate victim.

The skull of the early tyrannosauroid *Proceratosaurus* from the Middle Jurassic of the UK. In common with the majority of other theropod dinosaurs it was a predator whose sharp, curved teeth were ideally suited to slicing flesh.

Although the skulls of all theropods were built in similar ways, differences in their proportions, such as the length, width or height of the snout, or the connections between individual skull bones, resulted in diverse feeding adaptations. The shapes and sizes of the jaw muscles also varied, as did the directions they pulled in. For example, *Tyrannosaurus* had a skull that could withstand massive forces, with many of the individual skull bones being thickened and joined firmly together, making the whole structure incredibly strong. *Tyrannosaurus* also had massive jaw muscles, and calculations of its bite strength suggest that its jaws could exert forces of around 48,000 Newtons – about three times the bite force generated by a large saltwater crocodile (which has the strongest bite force measured for any living animal) – a truly bone-crushing bite. By contrast, tiny *Compsognathus* had a very open, lightly built skull with a maximum bite force of 15 Newtons (similar to a weasel) and was capable of crunching through only the smallest prey items.

# Herbivory and gut bacteria
## Hadrosaur jaw

PLANTS ARE ABUNDANT AND widespread, making them an ideal food source in many ways. They also have the distinct advantage of not running away when you try to eat them, although they do have some defences, such as thorns and spines. However, although plants look like an easy dinner option, it is incredibly difficult to unlock the energy stored within them. This is because leaves, stems, wood and fruits are composed primarily of two complex chemicals, cellulose and lignin, which are incredibly difficult to digest. In fact, all land animals lack the ability to break these compounds down. Instead, herbivores rely on help from an unexpected source – bacteria.

Some bacteria possess the chemical tools needed to break down cellulose, and herbivores survive by carrying these beneficial bacteria inside their guts. The herbivore provides a home for the bacteria and in return the bacteria break the cellulose down into simpler sugars that the animal can absorb through its gut. To accommodate the billions of bacteria needed, plant-eaters often have long digestive systems, sometimes

with specialized compartments for the bacteria to do their work. Dinosaurs likely did the same and these long guts are probably the main reason that many herbivorous dinosaurs reached large sizes, as the longer the gut the more efficient digestion becomes, allowing even the toughest plants to be eaten. Herbivores like elephants, rhinos and hippos are the largest land animals today for the same reason.

Chewing food thoroughly helps the bacteria to do their work, as pulped vegetation has a larger area for them to digest than the original uncrushed food. Most herbivorous dinosaurs had teeth that were leaf-shaped, with edges lined with large serrations, which tore through vegetation, so that it was at least partly chopped when swallowed. In many herbivores, like sauropods and early ornithischians, the teeth became closely aligned, slicing past each other precisely to work like shears. In others, like ornithopods and ceratopsians, the edges of the teeth rubbed against each other even more tightly, trapping food between them so that the food was chewed much more thoroughly. These dinosaurs had fleshy cheeks to stop food falling from the sides of the mouth as it was chewed. Many plant-eaters close their jaws in a simple scissor-like fashion, similar to carnivores, but ornithopods, ceratopsians and some ankylosaurs were able to move their lower jaws in more complex back-and-forth ways that allowed the teeth to grind food to a fine pulp. Hadrosaurs were arguably the most sophisticated plant eaters of all time, combining grinding movements of the lower jaw, strong jaw muscles, and a complex dentition. They were able to chew food just as thoroughly as any cow or sheep.

Lower jaw of a small hadrosaur from the Late Cretaceous of the USA, showing multiple tightly packed columns of teeth.

# Diet

## *Gorgosaurus*

DIRECT EVIDENCE OF DIET is usually lacking – so far, no skeleton has been found caught in the act of eating. Instead, palaeontologists rely on clues from tooth and skull shape to deduce what a dinosaur might have eaten. Just occasionally, however, some fossils give a much clearer picture of what was on the menu. A few skeletons include the fossilized remains of gut contents deep within their rib cages. Famous examples include skeletons of tiny lizards within the bodies of *Sinosauropteryx* and *Compsognathus*, a small crocodile skeleton in *Coelophysis*, as well as fish scales in *Baryonyx*, and the remains of two small *Oviraptor*-like theropods that formed the last meal of a *Gorgosaurus*. Carnivore gut contents are found more frequently than those of herbivores, because the bones and teeth of their prey are more likely to fossilize than mashed-up plant remains. Even so, a few herbivore gut contents are known, mainly for armoured dinosaurs, including fruits in the stomachs of *Kunbarrasaurus* and *Isaberrysaura* and fern leaves in the exceptionally preserved skeleton of *Borealopelta*. It is worth bearing in mind, however, that these last meals might not have been the animal's usual or entire diet – they are a single snapshot of what that individual dinosaur ate on the day it died.

Another important, but less savoury, source of information comes from dinosaur dung. A fossilized poo is referred to as a 'coprolite' and when examined in detail often contains useful clues on diet, such as bone or plant fragments. Unfortunately, we cannot usually match a coprolite to a particular dinosaur species, as skeletons and coprolites are rarely found together. One obvious exception comes from a couple of massive coprolites from the latest Cretaceous of Canada. These were clearly produced by a meat-eater, as they are filled with the crushed bones of duck-billed dinosaurs. At up to 67 cm (2¼ ft) long, the only theropod living in this time and place large enough to produce these coprolites was, of course, *Tyrannosaurus*. Once again, most dinosaur coprolites belong to carnivores: the high mineral content of their diet means that it is easier for them to turn into rock than the soft, mushy remains produced by herbivores. However, a few coprolites were left behind by plant-eaters, with the best examples containing chewed wood fragments; these were probably produced by hadrosaurs. Finally, bones themselves can give the game away: tooth scratches and puncture marks left on skeletons show where a carnivore feasted on its prey. Some of these confirm long-held suspicions, like the deep bite marks left on a *Triceratops* hip by a hungry *Tyrannosaurus*, as well as a few surprises, such as likely cannibalism in the large Madagascan theropod *Majungasaurus*.

This is a *Tyrannosaurus* coprolite - the largest piece of fossil dung from any dinosaur. It contains crunched up bone fragments from its unfortunate prey.

A skeleton of the tyrannosaurid *Gorgosaurus*. The area highlighted by the red box contains the bones of two other smaller theropod dinosaurs that formed its last meals.

# Brain size
## *Iguanodon*

ONE OF THE KEYS TO understanding an animal's behaviour comes from the size and architecture of its brain and nervous system. Brain size, and the proportions of its component parts, reflect the complexity and diversity of behaviour, as well as the relative importance of the different sense organs and their potential contributions to the animal's lifestyle. When discussing brain size, we are not interested in its weight or volume per se as, for obvious reasons, larger animals have bigger brains merely by dint of their greater size – and big farm animals are not noted for their crossword-solving skills. Instead, it is the mass of the brain relative to the animal's overall body mass that is important. There is a tight relationship between brain and body size in living animals, and it is usually possible to predict one from the other. The interesting cases occur where this relationship breaks down and the brain is either much larger or much smaller than expected based on body mass alone. For example, humans have brains that are about seven times larger than would be expected for their body mass and dolphin brains are four to five times larger than predicted, both findings that are consistent with complex social behaviours and high intelligence. By contrast, manatees and sloths have brains that are smaller than predicted, in accordance with their more sedentary habits.

Luckily, the sizes and shapes of dinosaur brains can be reconstructed with some confidence, as the interiors of their skulls accurately capture these features, as well as the courses of the major nerves leaving the brain and the blood vessels that supplied the brain with food and oxygen. The space formerly occupied by these structures can either be cast using rubber, to make a physical replica of the brain's shape or, nowadays, using computed tomography scans to make virtual models instead. These brain reconstructions are called endocasts. Occasionally, natural endocasts are formed where sediment fills the brain cavity during fossilization, capturing the shape of the brain. These are revealed when the surrounding bone is weathered away or carefully removed in the laboratory, as in the case of this *Iguanodon* skull, where the overlying bone was chipped away to reveal a perfect endocast along with inner ear structures, giving us a fantastic view of a 'real' dinosaur brain.

OPPOSITE: Sediment filled the brain cavity of this *Iguanodon* and the bone around it has been removed to reveal a cast of the brain, inner ear and major nerves.

*Iguanodon*'s brain was about the size we would expect for a dinosaur of its size. This is not to say that this endocast was not interesting or useful – it is just the standard size. As humans, we often regard intelligence as special and important, but it is worth bearing in mind that all a brain really has to do is coordinate its owner's ability to find food, get around, sense its environment, avoid danger and find a mate. Any animal that can do these things is doing just fine in an evolutionary sense. Most non-bird dinosaurs had brains that were either around the 'standard' size or smaller than body mass predicts, which is not too surprising given the massive sizes achieved by sauropods and others. However, as these groups were incredibly successful for millions of years it is clear that a large brain is not vital to thrive. Similarly, elephants, which have very complex social interactions, also have smaller brains than we would expect, because they have also reached much larger sizes than most mammals. By contrast, a few dinosaurs, mainly small-bodied theropods, had brains larger than we would predict, and this trend towards increased relative brain size continued into birds, some of which have brains six or more times the size expected. Although 'bird-brained' is often used as an insult, some birds, like crows, display advanced problem-solving skills and can use simple tools, which are more complex behaviours than seen in many mammals. Some dinosaurs were (and are) genuinely smart.

As different areas of the brain are responsible for vision, smell, coordination and behaviour, we can look at the overall shape of the endocast and the sizes of the different brain regions to gain insights into dinosaur biology. Many ornithischians, sauropods and early theropods had rather crocodile-like brains, with the key regions and nerves aligned in a straight line and rather small cerebral hemispheres (the parts of the brain associated with complex behaviours, which in humans make up the bulk of our brains). Coelurosaurs, and their avian descendants, have rather different looking, more globular, swollen brains, with large, expanded cerebral hemispheres. In some dinosaurs, such as the small herbivore *Leaellynasaura*, the areas of the brain associated with processing visual signals, the optic lobes, are relatively large, suggesting either acute vision or activity in low-light conditions (e.g. nocturnality). In others, like *Tyrannosaurus*, the olfactory lobes, those areas associated with smell, are expanded – perhaps indicating prowess in locating prey. A few dinosaurs have a short, sausage-like process that sticks out from the side of the brain, close to the inner ear. This feature, called the flocculus, is thought to be associated with rapid head movements and the ability to keep an animal's gaze fixed on a particular point in space even while it is moving around. A large flocculus seems to be of use in some predators (for keeping prey in view), some prey (for helping maintain head posture while running away) and birds (for coordinating movements through the air). The study of fossil brains and nervous systems is called palaeoneurology and computed tomographic scanning has enabled palaeontologists to reconstruct the endocasts of dozens of dinosaurs, allowing them to compare and document the many differences between them.

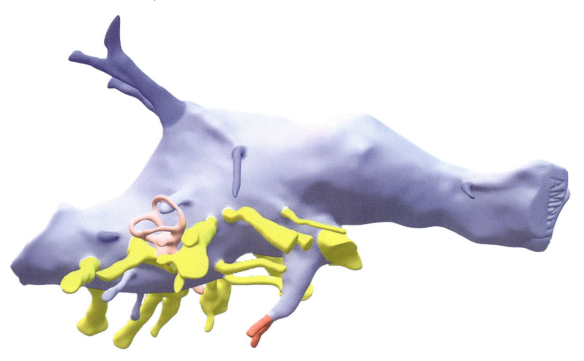

This is an endocast of *Tyrannosaurus rex*, based on CT scans of a well-preserved skull. The brain itself is shown in light blue, while the main nerves are shown in yellow and the inner ear in pink. The large tubular part of the brain on the right-hand side of the image represents the enlarged olfactory tract, responsible for the sense of smell.

# Hearing and sight
## *Haplocheirus*

WE EXPERIENCE THE WORLD through our senses: touch, hearing, sight, taste and smell. Some sense organs are built only of soft tissues, such as the taste receptors on our tongues or the pressure sensitive nerve endings in our fingertips. Sadly, these leave little (if any) trace on fossils. However, two senses – vision and hearing – are intimately associated with the skeleton. As a result, they leave clues that palaeontologists can interpret long after the relevant soft tissues have rotted away, helping to show us how dinosaurs perceived their world.

The size, shape and orientation of the eye socket (an opening called the orbit) gives us important information on the maximum size of the eyeball and the directions in which the eyes could see. Eyeball and pupil size are also revealed by a donut-shaped ring of tightly connected, delicate bony plates, called the sclerotic ring, which sits within the eyeball (sclerotic rings are present in birds and many reptiles, but are lacking in mammals). Among living animals, large eyes are associated with either sharp or low-light vision, or predatory behaviour; large pupil size is related to nocturnal habits. When we measure dinosaur sclerotic rings, some indicate huge pupils and large eyeballs, such as those belonging to the small Jurassic predator *Haplocheirus*. These provide strong evidence that *Haplocheirus* was nocturnal. Applying this method more widely, it seems that some other small theropods, including *Shuvuuia* and *Megapnosaurus*, were active at night, suggesting that this lifestyle was important for these dinosaurs at least. Other dinosaurs, like *Diplodocus* and *Psittacosaurus*, lacked these specializations and had eyes better suited for daytime use.

The behaviours revealed by eye size are backed up by studies on ear anatomy. Although we tend to think of ears as those fleshy flaps on the sides of our heads (though only mammals have these appendages), the most important parts of the ear are tucked away, out of sight. In particular, the organ responsible for hearing, the cochlea, is housed in a tight-fitting capsule buried within the skull wall. As it is completely encased by bone, its dimensions and shape can be extracted from fossils. Studies on living birds and reptiles show that the length of the cochlea is related to the range of sounds that an animal can hear: in general, the longer the cochlea, the better its hearing. The cochleae of *Haplocheirus* and *Shuvuuia* were enormous, comparable in length to those of our most familiar nocturnal hunters, owls. In contrast, the cochleae of the earliest bird, *Archaeopteryx*, were short, indicating that the *Archaeopteryx* ear was not tuned to hear high-pitched melodies, like those of modern songbirds, but was adapted for detecting low-pitched calls, similar to those made by crows.

A complete skull of the alvarezsaurid theropod *Haplocheirus*, from the Late Jurassic of China. In the original fossil, the sclerotic ring is present but has moved slightly out of position to lie in the bottom half of the eye socket.

CT-scanned skull of *Haplocheirus* showing the restored sclerotic ring (in blue) filling the eye socket. It also shows the inner ear (in pink) which is embedded within the side wall of the skull.

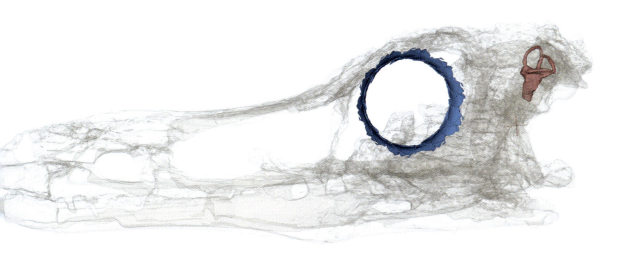

# Social behaviour
## *Parasaurolophus*

LIVING REPTILES ARE NOT generally thought to lead exciting social lives, but this is unfair as many species do engage in interesting (if underappreciated) behaviours – particularly when seeking mates. By contrast, most birds have rich, and sometimes spectacular, behavioural repertoires that include songs, calls, displays, gift-giving, dance and other interactions. For decades, palaeontologists assumed that dinosaurs were reptile-like in this regard, but a whole raft of discoveries now show that they engaged in complex social behaviours like those of birds and mammals.

A revealing illustration of these changing ideas comes from considering the crests of duck-billed dinosaurs. Without doubt, the most graceful of these belongs to *Parasaurolophus* from the Late Cretaceous of North America, whose elegant, backwardly curved, tubular crest can be up to 1.5 m (5 ft) in length. The crest is hollow and its tube-like interior is connected to the nose. For this reason, some earlier palaeontologists suggested that the crest might have been used as a snorkel, allowing *Parasaurolophus* to feed on aquatic plants (although it was pointed out that this could not have worked, as there is no breathing hole at its tip). The crest has also been viewed as a potential weapon, for defence, and as a tool for pushing the head through dense vegetation. However, it is now thought to have played an important role in display and communication instead. Experiments with model *Parasaurolophus* crests (both real and virtual) show that when air is pushed through the crest, via its connection with the nose, it helps to produce deep, booming notes, similar to those made by a trombone. These loud, low-frequency sounds would have been able to travel large distances and might have been useful in finding mates, staying together or signalling danger. In addition to sound, the size and distinctive shape of the crest suggests that it would have been used for showing off to others in displays over mating rights or territory. It has been proposed that a sail-like flap of skin connected the crest with the back of the skull, which could also have been used for display, but the presence of this structure is considered uncertain.

Many other hadrosaurs sport crests (but not all). Some of these crests are hollow, like those of *Parasaurolophus*, and their different lengths, shapes and internal piping might have created quite different types of species-specific calls. Other crests were solid

A beautiful skull of the hadrosaur *Parasaurolophus* showing just how long the crest could be in comparison with the rest of the head.

and for show only. Differently shaped crests characterized different species, from the domed, helmet-like crest of *Corythosaurus*, to the hatchet-shaped ornament seen in *Lambeosaurus* and the simple spike that adorns the skull of *Saurolophus*. The fact that these crests differ from species-to-species, and that young hadrosaurs had smaller or less complex versions of the crests present in adults, point to their social functions, implying that adults were using clues from their distinctive crest shapes for display, perhaps when selecting mates. This echoes examples from living animals, in particular deer and antelope, where many different antler and horn types are present, helping to distinguish species from each other and functioning in display, defence and fighting.

OPPOSITE: This Late Cretaceous scene shows hadrosaurids with different head crests, including *Parasaurolophus* (left), *Lambeosaurus* (right midground) and *Corythosaurus* (right foreground), which gave each of them a distinct appearance. Not all hadrosaurids had crests and in these cases palaeontologists have one less clue to use in deducing the social lives of these crestless species.

Similar comments apply to ceratopsians. Ceratopsian species are usually distinguished from each other by clear differences in skull shape, with variation in the overall size and shape of the frill, the shapes and numbers of the small bones and spikes decorating the frill's margins and, of course, the numbers, positions and sizes of the horns (situated over the eyes or on the nose) being the most important characteristics. Anyone familiar with the great herds of antelope on the plains of Africa today will immediately see the similarities between the mixture of horns among wildebeest, eland and kudu, and those of *Chasmosaurus*, *Styracosaurus* and *Centrosaurus*. Although it used to be thought that horns and frills were primarily for defence, as shown in myriad reconstructions of iconic battles between *Triceratops* and *Tyrannosaurus*, it is much more likely that they served a social purpose. This is because extensive studies on antelope, deer and other living animals with similar weaponry show that they use their horns and antlers mainly for fighting each other, in contests over access to mates and resources. Although they are also of use in fending off the attentions of hungry predators, they are used for defence more rarely.

Exceptional fossil sites provide other evidence of social behaviour. For example, most tracksites record very short time periods, such as an interval between high tides. Consequently, discoveries of multiple, tightly bunched trackways, with many animals moving in the same direction, are good evidence of herding: all these animals were in the same place at the same time. In addition, some rock layers are so rich in skeletons that they are termed 'bonebeds'. These may contain anything from a handful of tightly entwined individuals to hundreds (perhaps thousands) of specimens. One type of bonebed is formed in an instant, as the result of a mass death event. These die-offs are most often caused by flash floods (leading to instant burial) or droughts (with groups of animals dying of thirst at the same time), but are sometimes formed by volcanic ash flows. These 'catastrophic bonebeds' capture populations of animals that were living together at the moment of their death. Most bonebeds of this type are dominated by the remains of a single species, providing unambiguous proof of herding. Palaeontologists have now found abundant evidence for group-living in numerous dinosaur species, mainly for plant-eating ornithischians and sauropods, but sometimes among theropods too.

# A global phenomenon
## Unnamed ornithopod

IN POPULAR CULTURE, DINOSAURS are frequently miscast as dull, stupid, unsuccessful animals. Indeed, the very word 'dinosaur' is often used as an insult, implying that something is outdated or doomed to failure. Nothing could be further from the truth – and we should be using this label as a compliment instead. Despite our accomplishments (and mistakes) as a species, and our vast numbers, *Homo sapiens* has walked the planet for only 300,000 years and been ecologically important for just the last few thousand of those. By contrast, non-bird dinosaurs occupied almost all the major ecological roles on land for nearly 170 million years: a period nearly 600 times longer. When we add in bird diversity the total dinosaur record stretches to nearly 240 million years, and they retain a global distribution, found almost everywhere on land, sea and in the sky, with over 11,000 living species. Dinosaurs are still very much with us and continue to play important roles in modern ecosystems.

During the Mesozoic Era, dinosaurs expanded from humble beginnings as rare members of early Late Triassic ecosystems to achieve a vast distribution that stretched from pole to pole in all conceivable habitats, which ranged from lush forests to deserts. This lower jaw, part of the skeleton of an as-yet undescribed species, exemplifies their adaptability. It belongs to a medium-sized ornithopod dinosaur from Antarctica, collected from the ice-bound slopes of one of the islands dotting the seas around the

The lower jaw of an unnamed ornithopod dinosaur from Antarctica – one of many new species being uncovered by museum palaeontologists.

Antarctic Peninsula. Antarctica was not an icy desert during the Late Cretaceous, due to the greenhouse-like conditions of the Cretaceous atmosphere, and this ornithopod would have feasted in the polar forests that then clothed the region. Nevertheless, Antarctica at this time was not too far north of its current position, meaning that this dinosaur dealt with months of polar darkness and the low temperatures of the polar winter. So far, only a handful of dinosaur fossils have been recovered from the continent, due to its harsh current conditions, but the few we have already show that a range of species was present, even here. Discoveries at the other end of the world, from the North Slope of Alaska, have also revealed dinosaurs that endured frigid winter temperatures – imagine herds of hadrosaurs battling through blizzards as they were stalked by dromaeosaurs whose feathers were encrusted in ice.

Dinosaurs were unqualified successes: they achieved a size range on land unrivalled by any other group; had diverse diets exploiting the widest variety of plants and prey; they took to the air; their numbers included runners, diggers, gliders and climbers; they sported some of the most impressive and bizarre ornaments; and evolved complex social behaviours. However, with the notable exception of birds, none of their diverse adaptations, vast numbers or wide distributions helped them when catastrophe struck.

# End of an era
## *Tyrannosaurus rex*

SIXTY-SIX MILLION YEARS AGO, a cataclysm brought the Cretaceous Period, and the Mesozoic Era, to an end. It led to the extinction of all non-bird dinosaurs, as well as around 70 per cent of all other animal life on land and in the oceans. Nothing that weighed more than 25 kg (55 lb) survived on land. The causes of this mass extinction, the second greatest in Earth's history, have been debated intensively, and various culprits have been suggested, ranging from titanic volcanic eruptions, to tiny mammals over-indulging on dinosaur eggs, to the explosion of a nearby star. However, palaeontologists now think that the primary cause of this calamity was the unexpected arrival of a giant rock from space.

In the late 1970s, geologists working in the Italian Alps noted high concentrations of a rare chemical element in rocks laid down at the very end of the Cretaceous. This element, iridium, is exceptionally rare on Earth but abundant in meteorites. Probing further, similar samples of the same age from all over the world were soon found to be rich in iridium. The recognition of this widespread chemical anomaly led quickly to the idea that a gigantic meteorite must have struck the Earth at this time, enriching the rocks with iridium and causing the extinction. Palaeontologists

Even the mighty *Tyrannosaurus* was unable to survive the catastrophic events that ended the Cretaceous Period.

were initially sceptical, as no impact crater of sufficient size or the right age was then known. This was to change just a few years later, when a vast buried crater was discovered on the seabed off the coast of the Yucatan Peninsula, Mexico. Rock samples drilled from this ring-like structure, which is 180 km (110 miles) in diameter, were dated to the very end of the Cretaceous. Analysis of the crater now indicates that it was formed by the impact of an asteroid that was 10 km (6 miles) in length and travelling at close to a staggering 72,400 kmph (45,000 mph). More evidence has since accumulated in support of this asteroid impact hypothesis. The impact caused earthquakes and gigantic tidal waves, and it blasted a devastating heatwave around the world. It sent millions of tonnes of liquified rock into the air, and wildfires added vast amounts of soot and ash to the atmosphere. By unlucky coincidence, the asteroid hit an area of sulphur-rich rocks, creating clouds of acid rain. All these changes occurred within just a few hours, on a global scale. Those animals and plants closest to the impact were killed immediately; others were killed by its consequences. Plant life was devastated as sunlight was blocked out for years. As plants died, the herbivores dependent on those plants starved; in turn, as herbivores vanished, predators died. Food webs collapsed catastrophically. Among dinosaurs, birds had the advantage of being small, and of having flexible diets and the ability to fly long distances in search of better prospects: some escaped the worst effects of the disaster, eventually giving rise to the birds around us today. By contrast, food shortages signed the death warrants of their larger, less mobile cousins.

Although a few palaeontologists have suggested that dinosaurs were in decline before the extinction, due to long-term changes in climate and geography, most agree that dinosaurs were still thriving at the time of the asteroid impact. The last surviving non-bird dinosaurs include icons like *Tyrannosaurus*, *Triceratops*, *Edmontosaurus* and *Ankylosaurus*, and it is humbling to think of these animals gazing up at the sky and seeing a new, unfamiliar flaming ball of light, having no conception of the havoc that was about to be unleashed. It is possible that a few non-bird dinosaur species clung on to life for a few hundred years after the impact, particularly in areas far removed

from the Gulf of Mexico, but it is notable that we have never found any fossil in post-Cretaceous rocks to suggest that this was the case.

The story of dinosaur evolution is bracketed by disasters, with the end-Triassic mass extinction instrumental in their initial rise and the end-Cretaceous extinction snuffing out all dinosaurs but birds. Despite their success, dinosaurs were wiped out in an instant – they had no way of adapting to the instantaneous, massive changes brought by the asteroid. Their extinction was simply bad luck. Sadly, our living dinosaurs are under threat too, but from a quite different source – us. Humans, with their hunger for land and resources are making life on our planet intolerable for many of the species we share it with, and biologists have labelled the present rate of species loss 'the sixth mass extinction'. Worryingly, this would be the first mass extinction caused not by asteroids, volcanoes or other natural changes, but by the actions of one dominant species. However, it is well within our power to prevent living dinosaurs from suffering yet another extinction if we act promptly to conserve the habitats on which they depend.

Studying dinosaurs shines a light onto what is biologically possible: they pushed at the extremes of what evolution can achieve with blood, muscle and bone. They show how their living representatives evolved and provide us with a unique, exciting experiment in the history of life. Just as importantly, they illustrate how even the most successful denizens of the planet can be wiped out at the very peak of their prowess. We can learn lessons from this. Humanity is just as much a part of the natural world as dinosaurs were, and we need to respect it, lest we hurry our own way into oblivion.

Despite its fame, *Tyrannosaurus rex* was one of the last non-bird dinosaurs to evolve and the species lived for only 2–3 million years before it, and its contemporaries, were wiped out by the asteroid impact.

# Geological timescale

| EON | ERA | PERIOD OR EPOCH | AGE (millions of years) |
|---|---|---|---|
| PHANEROZOIC | CENOZOIC | Holocene | 0.012 |
| | | Pleistocene | 2.6 |
| | | Pliocene | 5.3 |
| | | Miocene | 23 |
| | | Oligocene | 34 |
| | | Eocene | 56 |
| | | Paleocene | 66 |
| | MESOZOIC | Cretaceous | 145 |
| | | Jurassic | 201 |
| | | Triassic | 252 |
| | PALAEOZOIC | Permian | 299 |
| | | Carboniferous | 359 |
| | | Devonian | 419 |
| | | Silurian | 443 |
| | | Ordovician | 485 |
| | | Cambrian | 541 |
| PRECAMBRIAN | | Ediacaran | 635 |
| | | Cryogenian | 850 |
| | | | 4600 |

1. The vertical (time) axis is not to scale
2. Only the two youngest periods and none of the eras of the Precambrian are shown
3. Epochs rather than periods are specified for the Cenozoic

pp.8–9
*Megalosaurus bucklandii*
(original lower jaw)
Middle Jurassic;
Oxfordshire, England,
UK
OUMNH J.13505

pp.12–13
*Mantellisaurus
atherfieldensis* (original
near-complete skeleton)
Early Cretaceous; Isle of
Wight, England, UK
NHMUK PV R5764

p.17
*Teleocrater rhadinus*
(original ilium)
Middle Triassic; Ruhuhu
Basin, Tanzania
NHMUK PV R6795

p.21
*Nyasasaurus parringtoni*
(original humerus and
sacral vertebrae)
Middle Triassic; Ruhuhu
Basin, Tanzania
NHMUK PV R6856

pp.22–23
*Herrerasaurus
ischigualastensis*
(original skull)
Late Triassic; San Juan
Province, Argentina
PVSJ 407

pp.24–25
*Eoraptor lunensis* (replica
complete skeleton)
Late Triassic; San Juan
Province, Argentina
(Replica based on PVSJ
512)

pp.26–27
*Eoraptor lunensis*
(original skull)
Late Triassic; San Juan
Province, Argentina
PVSJ 512

pp.28–29
*Coelophysis bauri* (replica
complete skeleton)
Late Triassic; New Mexico,
USA
NHMUK PV R8927
(replica of AMNH FR
7224)

p.31
*Plateosaurus trossingensis*
(original near-complete
skeleton)
Late Triassic; Aargau
Canton, Switzerland
MSF 23

p.33
*Lesothosaurus diagnosticus*
(original skull)
Early Jurassic; Quthing,
Lesotho
NHMUK PV RU B23

pp.36–37
*Massospondylus carinatus*
(original near-complete
skeleton)
Early Jurassic; Free State
Province, South Africa
BP/1/4934

p.39
*Heterodontosaurus tucki*
(original skull)
Early Jurassic; Eastern
Cape Province, South
Africa
AM 4766

pp.40–43
*Scolosaurus cutleri* (original
partial skeleton)
Late Cretaceous; Alberta,
Canada
NHMUK PV R5161

pp.44–45
*Stegosaurus stenops*
(original near-complete
skeleton)
Late Jurassic; Wyoming,
USA
NHMUK PV R36730

pp.46–47
*Pachycephalosaurus
wyomingensis* (replica
skull)
Late Cretaceous; Montana,
USA
NHMUK PV R10053
(replica of AMNH 1696)

pp.48–49
*Triceratops horridus*
(replica skeleton)
Late Cretaceous;
Wyoming, USA
NHMUK PV R10957
(composite replica of
specimens in USNM)

153

SPECIMEN DETAILS

pp.50–51
*Triceratops* sp. (species uncertain) (original horn core)
Late Cretaceous; Wyoming, USA
NHMUK PV R3677

pp.52–53
*Hypsilophodon foxii* (original skull)
Early Cretaceous; Isle of Wight, England, UK
NHMUK PV R197

pp.54–55
*Hypsilophodon foxii* (original near-complete skeleton)
Early Cretaceous; Isle of Wight, England, UK
NHMUK PV R5830

pp.56–57
*Edmontosaurus regalis* (original near-complete skeleton and close up on original skull)
Late Cretaceous; Alberta, Canada
NHMUK PV R8927

p.59
*Pantydraco caducus* (original skull and neck)
Late Triassic; Vale of Glamorgan, Wales
NHMUK PV P24

pp.64–65
*Diplodocus carnegii* (replica skeleton)
Late Jurassic; Wyoming, USA
NHMUK PV R8642 (composite replica based on specimens in CMNH and USNM)

p.67
*Giraffatitan brancai* (original composite skeleton)
Late Jurassic; Tendaguru, Tanzania
MB.R.2181

p.69
*Giraffatitan brancai* (original hand)
Late Jurassic; Tendaguru, Tanzania
MB.R.2181.79.1–11

pp.70–71
*Dilophosaurus wetherilli* (original partial skull)
Early Jurassic; Arizona, USA
UCMP 77270

p.73
*Carnotaurus sastrei* (original skull)
Early Cretaceous; Argentina
MACN CH 894

pp.74–75
*Spinosaurus aegypticus* (replica skeleton)
Late Cretaceous; Errachidia, Morocco
Replica in FMNH (based on FSAC-KK 11888)

pp.78–79
*Allosaurus fragilis* (replica skull)
Late Jurassic; Utah, USA
NHMUK PV R12153 (replica based on specimens in UMNH)

p.81, p.83
*Sinosauropteryx prima* (original complete skeleton, with close-up on skull)
Early Cretaceous; Liaoning Province, China
NGMC V2123

pp.84–85, p.87
*Deinonychus antirrhopus* (replica skeleton and foot)
Early Cretaceous; Montana, USA
Replica skeleton in HMNS (based on various fossils in YPM including YPM 5205); foot in Museum of the Rockies, Montana collected Carbon County, Montana

p.89
*Archaeopteryx lithographica* (original near-complete skeleton)
Late Jurassic; Bavaria, Germany
NHMUK PV OR37001

p.91
*Confuciusornis sanctus* (original complete skeleton)
Early Cretaceous; Liaoning, China
NGMC V2130

pp.92–93
*Haestasaurus becklesii*
(original skin
impression)
Early Cretaceous; East
Sussex, England, UK
NHMUK PV R1868

p.95
*Sinornithosaurus millenii*
(original complete
skeleton)
Early Cretaceous;
Liaoning, China
NGMC V91

p.99
*Anchiornis huxleyi*
(original complete
skeleton)
Late Jurassic; Liaoning,
China
LPM-B00169A

p.101
*Scipionyx samniticus*
(original nearly complete
skeleton)
Early Cretaceous;
Benevento, Italy
SBA-SA 163760

p.103, pp.104–105
Titanosaur egg, genus
and species unknown
(original)
Late Cretaceous; Madhya
Pradesh, India
NHMUK BM.58644

pp.110–111
Unnamed oviraptorosaurid
(original partial skeleton
with eggs)
Late Cretaceous; Jiangxi
Province, China
ZMNH M8829

p.115
*Patagotitan mayorum*
(replica skeleton)
Early Cretaceous; Chubut
Province, Argentina
(Replica based on MEF-
PV 3400 and others)

p.117, pp.118–119
*Ngwevu intloko*
(original bone cross-
section and skull)
Early Jurassic; Free State
Province, South Africa
BP/1/4779

p.121
*Diluvicursor pickeringi*
(original partial skeleton)
Early Cretaceous; Victoria,
Australia
NMV P221080

p.123
*Ornithopsis hulkei*
(original vertebra)
Early Cretaceous; Isle of
Wight, England, UK
NHMUK PV R89

p.125
Dinosaur trackway,
possibly *Megalosaurus
bucklandii* (original)
Middle Jurassic;
Oxfordshire, England, UK

p.127
Dinosaur track, probably
an ornithopod
(original)
Middle Jurassic; Isle of
Skye, Scotland, UK

p.129
*Tuojiangosaurus
multispinus* (original
restored skeleton)
Late Jurassic; Sichuan
Province, China
CV 209

pp.130–131
*Proceratosaurus bradleyi*
(original partial skull)
Middle Jurassic;
Gloucestershire,
England, UK
NHMUK PV R4860

pp.132–133
Unnamed hadrosaurid
(original lower jaw)
Late Cretaceous; Alberta,
Canada
NHMUK PV R4472

p.135 (top)
*Tyrannosaurus rex*
coprolite (original)
Late Cretaceous;
Saskatchewan, Canada
SMNH P.2609.1

p.135 (bottom)
*Gorgosaurus libratus*
 (original partial skeleton with gut contents)
Late Cretaceous; Alberta, Canada
TMP 2009.12.14

p.137
*Iguanodon bernissartensis*
 (original partial skull with natural endocast)
Early Cretaceous; West Sussex, England, UK
NHMUK PV R8306

p.141
*Haplocheirus sollers*
 (original skull)
Late Jurassic; Xinjiang, China
IVPP V14988

pp.142–143
*Parasaurolophus walkeri*
 (replica skull)
Late Cretaceous; Alberta, Canada
NHMUK PV R4859
 (replica of ROM 768)

pp.146–147
Unnamed ornithopod
 (original lower jaw)
Late Cretaceous; Vega Island, Antarctica
NHMUK PV R36760

pp.148–149
*Tyrannosaurus rex*
 (original complete skeleton)
Late Cretaceous; South Dakota, USA
FMNH PR2081

pp.150–151
*Tyrannosaurus rex*
 (original lower jaw)
Late Cretaceous; Wyoming, USA
NHMUK PV R8020

AM, Albany Museum, Makanda, South Africa; AMNH, American Museum of Natural History, New York, USA; BP/1, Evolutionary Studies Institute, University of the Witwatersrand, Johannesburg, South Africa; CMNH, Carnegie Museum of Natural History, Pittsburgh, USA; CV, Chongqing Museum of Natural History, Chongqing, China; FMNH, Field Museum, Chicago, USA; FSAC, Faculté des Sciences Aïn Chock, Casablanca, Morocco; HMNS, Houston Museum of Natural Science, Houston, USA; IVPP, Institute of Vertebrate Paleontology and Paleoanthropology, Beijing, China; MACN, Museo Argentino de Ciencias Naturales 'Bernardino Rivadavia', Buenos Aires, Argentina; MB, Museum für Naturkunde, Berlin, Germany; EF, Museo Paleontológico Egidio Feruglio, Trelew, Argentina; MSF, Sauriermuseum Frick, Frick, Switzerland; NGMC, Geological Museum of China, Beijing, China; NIGP, Nanjing Institute of Geology and Paleontology, Nanjing, China; NHMUK, Natural History Museum, London, UK; NMV, Museum Victoria, Melbourne, Australia; OUMNH, Oxford University Museum of Natural History, Oxford, UK; PVSJ, Museo de Ciencias Naturales, Universidad Nacional de San Juan, San Juan, Argentina; ROM, Royal Ontario Museum, Toronto, Canada; SBA, Soprintendenza per i Beni Archeologici di Salerno, Avellino, Italy; SMNH, Royal Saskatchewan Museum, Regina, Canada; TMP, Royal Tyrrell Museum of Palaeontology, Drumheller, Canada; UCMP, Museum of Paleontology, University of California Berkeley, Berkeley, USA; UMNH, Utah Museum of Natural History, Salt Lake City, USA; USNM, National Museum of Natural History, Smithsonian Institution, Washington, DC, USA; ZMNH, Zhejiang Museum of Natural History, Hangzhou, China.

# Index

Page numbers in *italic* refer to illustration captions.

abelisaurids 72–3
abundance 30, 34–5, *37*
*Aepyornis* 102
aetosaurs 22
Africa 32, 35, 38, 41, 52, 54; *see also* Egypt; Lesotho; Morocco; South Africa; Tanzania *and* Zimbabwe
agate 102, *104*
age of dinosaurs *see* lifespan
air sacs 27, 61, 122, *123*
Alaska, USA 120, 147
*Allosaurus* 7, 78–9, 154
alvarezsaurids *141*
*Anchiornis* 98–9, 99, 114, 130, 155
ankle bones 18, 19
ankylosaurs 41, 43, 128, 133
*Ankylosaurus* 32, 41, 43, 150
Antarctica *31*, 43, 55, 58, 71, 120, *121*, *146*, 146–7, 156
*Apatosaurus* 119
apex predators 70, 72, 79, 130
aquatic dinosaurs 74–5, 77, 77
*Archaeopteryx* 88, *88*, 90, 98, 140, 154
archosaurs 16
Arctic 120
Ardley Quarry, UK 124, 126
Argentina 20, 22, 24, *24*, 27, *31*, 38, 72, 90, *115*, 153, 154, 155
*Argentinosaurus* 68, 114
Arizona, USA 154
armour 16
    and quadrupedality 128–9, *129*
    *see also* osteoderms; spikes; spines *and* vertical plates
armoured dinosaurs 40–5, 134
Asia 46, 51, 52, 55; *see also* China; India; Mongolia *and* Thailand
asteroid impacts 149–50
Australia 51, 77, 120, *121*, 155

back sails 77
Bahariya Oasis, Egypt 74, 75
*Baryonyx* 77, 134
beak 32, 48, 50, 56, 82, 90
bee hummingbird 114
*Beipiaosaurus* 98
Belgium *15*, 55
Bernissart, Belgium 55
bipedalism 14, 23, 26, 28, 30, 32, 38, 46, 48, 52, 124

birds
    early diversification 90, *91*
    evolution of flight 88
    feathers 94, *94*, 96–7, *97*
    origins 5, 84–7, 90, 94
bite *39*, *73*, 80, 131
body length
    up to 2 m (up to 6½ ft) 22, 56, 100, 114
    2 m up to 5 m (6½ ft up to 16½ ft) 18, 22, 29, 32, 38, 48, 50, 82, 85
    5 m up to 10 m (16½ ft up to 33 ft) *11*, 35, 43, 46, 68, 70, 72, 79, 85, 126
    10 m or more (33 ft or more) *31*, 51, 55, 56, 65, 66, 77, 79, 80
body mass 61, 96, 114, 119, 128, 136, 138
body size, extremes of 31, 60–1, 114
body temperature regulation 44, 77, 94, 97, 120
body weight
    up to 100 kg (up to 220 lb) 22, 29, 54, 41, 114
    100 kg up to 1 tonne (220 lb up to 1 ton) 22, 41, 50, 68, 82
    1 tonne up to 10 tonnes (1 ton up to 11 tons) 30, 43, 54, 56, 65, 77, 82, 128
    10 tonnes up to 20 tonnes (11 tons up to 22 tons) 51, 56, 61, 128
    20 tonnes or more (22 tons or more) 66, 68, 114
bone growth records 116, *117*
bonebeds 78, 145
*Borealopelta* 43, 98, 134
*Brachiosaurus* 60, 66
brain 80, 136, *136*, 138–9, *139*
Brazil 19, 20, 24, 30
breastbone 90
*Brontosaurus* 60
Buckland, William 9, 11
*Buriolestes* 30

*Caihong* 98
*Camarasaurus* 66, 68
camouflage 93, 98, 105
Canada 41, 43, 46, 56, 93, 134, 153, 154, 155, 156
cannabalism 134
*Carcharodontosaurus* 79, 130
carnivores *see* predatory dinosaurs

carnivory 14, 23, *23*, 25, 26, 28, 30, 77, 80, 134
*Carnotaurus* 72–3, *73*, 154
*Caudipteryx* 98
*Centrosaurus* 51, 145
ceratopsians 47, 48–51, 120, 128, 133, 145
*Ceratosaurus* 71
*Chasmosaurus* 145
chewing 38, 52, 133
Chile 43
China *31*, 38, 41, 44, 45, 48, 50, 56, 60, *61*, 71, 80, 90, 94, 97, *111*, 113, *129*, *141*, 154, 155, 156
*Citipati* 82, 109
classifications of dinosaurs 27
claws 23, 29, 30, 54, 79, 83, 84, *87*, 90, 100, 124, 126, 128, 130
Cleveland-Lloyd Dinosaur Quarry, USA 78
climbing 90
Cloverly Formation, USA 84, 85
coal 8
cochlea 140
*Coelophysis* 2, 28–9, *29*, 70, 78, 134, 153
coelurosaurs 80, 82–3, 97
cold survival 120, *121*, 147
colour 51, 70, 93, 97, 98, 99, 105
common ancestor 12, 14, 16, 26
communal nesting 106, 109
communication 33, 65, 142
*Compsognathus* 131, 134
*Confuciusornis* 90, *91*, 112, 154
cooperative hunting 85
coprolite 134, *135*
*Corythosaurus* 144, *145*
countershading 98
crests 33, 57, 70–1, *71*, 82, 142–4, *143*, 145
crocodile-like dinosaurs 16, 19
*Cryolophosaurus* 71
Cuba 114
Cuvier, Georges 11
cynodonts *18*, 22

Darwin, Charles *88*
defence 30, 38, 43, 44, 51, 54, 58, *64*, 65, 68, 106, 126, 142, 144–5
*Deinocheirus* 82
*Deinonychus* 84–7, *87*, 105, 120, 154
*Dicraeosaurus* 65

dicynodonts *18*, 22
diet
    early dinosaurs 14
    evidence of 134, *135*
    *see also* carnivory; herbivory; omnivory *and* piscivory
digestive system 100, 132–3; *see also* gut
*Dilophosaurus* 70–1, *71*, 154
*Diluvicursor* 120–1, *121*, 155
*Dimetrodon* 77
dinosaur characteristics 12–15
dinosaur groups, divergence of 5, 24–7
*Diplodocus* 7, 12, 24, 30, *64*, 64–5, 66, 68, 114, 140, 154
display 33, 38, 44, 47, 57, 71, 73, 77, 82, 90, 93, 97, *97*, 99, 112, 142–4
distribution 30, *31*, 35, 43, 52, 55, 56, 56, 58, 60, 61, 68, 146–7
divergence of dinosaur groups 5, 24–7
dome-headed dinosaurs *see* *Pachycephalosaurus*
dromaeosaurs 147
duck-billed dinosaurs 134; *see also* hadrosaurs
dung *see* coprolite

ear anatomy 136, *136*, *139*, 140, *141*
early dinosaurs 20–1, *21*, 22–3
ectothermy 120
*Edmontosaurus* 56–7, *56*, *57*, 93, 150, 154
egg-laying 83, 106, 112
Egg Mountain, USA 106, *107*
eggs 58
    brooding 94, 97, 109
    titanosaur 102, *102*, *104*, 104–5
    unlaid *111*, 113
    *see also* parental care
eggshells *102*, 104, 112, 121
Egypt 74, 75
embryos 58
endocasts 136, *136*, 138, *139*
endothermy 120, 121
*Eoabelisaurus* 72
*Eoraptor* 24, *24*, 27, *27*, 153
*Erlikosaurus* 82
Europe 45, 52, 54, 77; *see also* Belgium; Germany; Italy; Portugal; Spain; Switzerland *and* UK
'Evolutionary' theory *88*

evolutionary tree 5
extinction events
  end-Cretaceous 43, 148–51
  end-Triassic 34–5, 70, 151
  'sixth' 151
extinction of non-bird dinosaurs 148–51, *151*
eye anatomy 28, 32, 41, 140, *141*
eyesight 28

fat storage 77
feathered theropods 71, 80–3, 97, 120
feathers 88, 90, *91*, 94, *94*, 96–7, *97*, 112
feet 80, 86
  claws 84, *87*, 90, 124, 126, 130
  toes 28, 84, 124
  *see also* footprints
female dinosaurs *91*, 110, *111*, 112, 113
Field Museum, Chicago 110
filaments *83*, 94, 97
finite element analysis 78
fish-eating *see* piscivory
flight 16, 86, 88, 90
flocculus 138–9
footprints 43, 124, *125*, 126, *127*
fossil assemblages 78, 145
fossil discoveries, rate of 5
frills 33, 48, 50–1, 112, 128, 145
*Fruitadens* 114

Geological Society of London 11
geological timescale 152
Germany 30, 41, 66, 75, 88, 154
Ghost Ranch, USA 28, 78
*Giganotosaurus* 77
gigantothermy 121
*Gigantspinosaurus* 44
*Giraffatitan* 66, *66*, 68, *69*, 154
*Glacialisaurus* 58
*Gorgosaurus* 134, *135*
Greenland *31*, 58
growth rates 20, *109*, 116–19, *117*, *119*, 120
gut 100
  contents 28, 43, 77, 90, 134, *135*
  length 32, 54, 58, 128–9, 133
gut bacteria 132–3

habitats 58, 146
hadrosaurs 55, 56–7, 120, 126, 128, 134, 147
  crests 142–4, *143*, *145*

eggs 105
  lower jaw and teeth 132–3, *133*, 155
*Haestasaurus* 92–3, *93*, 155
handprints 124
hands 38, 52, 54–5, 60, *69*, 80
  claws 23, 29, 30, 79, 83, 100, 128, 130
  fingers 29, 30, 57, 73, 82, 86
  thumb 55
*Haplocheirus* 140, *141*
hatchlings 58, 106, *107*, 108
head size 30, 50, 80, 82
hearing *see* ear anatomy
Hell Creek Formation, USA 51
Hendrickson, Sue 110
herbivores 22, 35, 41, 43, 46, 51, 56, 58, 61, 68, 82, 134, 138, 150
herbivory 24, 26
  adaptations for 32–3, 52–5
  and gut bacteria 132–3
herding 30, 33, 35, 52, 58, 120, 126, 145, 147
*Herrerasaurus* 2, 22–3, *23*, 25, 153
heterodontosaurids 114
*Heterodontosaurus* 38–9, *39*, 153
hips 14, 32, *43*
  bones *16*, *18*, *19*, 27, 32, 86
  size 126
  width 43, 68
holotype specimens 53
*Homo sapiens* 146
horned dinosaurs *see* ceratopsians
horns 33, 38, 51, *51*, 72, 73, 112, 145
Huxley, Thomas Henry 84, 86
*Hypsilophodon* 52, *53*, 54, *55*, 154

*Ichthyovenator* 77
*Iguanodon* 26, 27, 55, 136, *136*
iguanodontians 54–5
India 19, 54, 102, 105, 155
intelligence 136
iridescence 98
iridium 148–9
*Irritator* 77
*Isaberrysaura* 134
Ischigualasto, Argentina *24*
island dwarfism 69
*Issi* 58
Italy 100, 155

Japan 43
jaw, lower *9*, 28, 32, 74, 132–3, *133*, 146, *146*; *see also* teeth

Jehol Biota, China 90, 94, 97
*Jurassic Park* (film) 70, 85
juveniles *33*, *47*, *55*, 119, 126, 144

*Kentrosaurus* 44
keratin 32, 51, *51*, 70, 72
*Kulindadromeus* 93, 97
*Kunbarrasaurus* 134

lagerpetids 16
*Lambeosaurus* 144, *145*
last meal *see* gut: contents
*Leaellynasaura* 138
length *see* body length
Lesotho 153
*Lesothosaurus* 32–3, *33*, 41, 118, 153
lifespan 116, 118
limb bones
  differing lengths 44, 46, 66
  growth records 116, *117*
  humerus 20, *21*
limestone rocks 8

*Macroelongatoolithus* 102
Madagascar 72, 102, 134
*Magyarosaurus* 68–9
*Maiasaura* 106–9
*Majungasaurus* 134
male dinosaurs 112
*Mamenchisaurus* 60–1, *61*
Mantell, Gideon *15*, *55*, 93
*Mantellisaurus* 12, *13*, *15*, 54, 55, 153
*Masiakasaurus* 72
*Massospondylus* 34, 35, *37*, 58, 119, 153
meat-eating *see* carnivory
medullary bone 112
*Megalosaurus* 8, 9, 11, *11*, *125*, 126, 153, 155
*Megapnosaurus* 140
melanosomes 98
metabolic rates *85*, 120–1
Mexico 149
*Microraptor* 98, 114, 130
Mongolia 48, 50, 82, 106, 108, 126
Monolophosaurus 71
Montana, USA *51*, 84, 106, *106*, 153, 154
Morocco 75, 154
Morrison Formation, USA *7*, 78
mortality rates 119
Museum for Nature, Berlin 66
*Mussaurus* 58

naming of dinosaurs 9, 11, 114
Natural History Museum, London 7, 44, 88, 102
neck 19, 29, *59*, 72

neck length 30, 58, 60–1, *61*, 66
*Nemegtomaia* 109
nerves *136*, *139*
nesting behaviour 106, 108–9, *109*
nests 58, 82, 106, *107*
New Mexico, USA 28, 78, 153
*Ngwevu* 116–19, *117*, *119*, 155
noasaurids 72
nocturnality 138, 140
non-bird dinosaurs
  extinction 148–51, *151*
  feathered *80*, 80–3, 94, 97
North America 48, 51, 52, 54, *56*, 72, 82, 142; *see also* Canada; Mexico and USA
North Atlantic Ocean 34
nostrils *57*, 65, 66
*Nyasasaurus* 20–1, 153

olfactory lobes 138
omnivory 14, 26, 38, 58, 130
optic lobes 138
origins
  birds 5, 84–7, 94
  dinosaurs 16–19
ornithischians 26, 27, 32, 35, 38, 39, 40, 47, 48, 97, 122, 133
ornithomimosaurs 82
ornithopods 52–5, *121*, *127*, 133, *146*, 146–7
*Ornithopsis* 122, 123, 155
Ornithoscelida 27
osteoderms *40*, 40–1, *42*, 44, 68
osteohistology 116, *117*
Ostrom, John 84, 86
*Oviraptor* 71, 82, 109, 134
oviraptorosaurs 82–3, 102, 109, 110, *111*, 113, 155
Owen, Richard 114
Oxford, University of 9, 11

*Pachycephalosaurus* 46–7, *47*, 153
*Pachyrhinosaurus* 51
palaeoart reconstructions *7*, *11*, *15*, *18*, *61*, *77*, *97*, *98*, *99*, *109*, *145*
palaeoneurology 139
palpebral 32
Pangaea 34
*Pantydraco* 58, *59*, 154
*Parasaurolophus* 32, 57, 142–3, *143*, 144–5, *145*
parental care 106, 108–9
*Parvicursor* 114
Patagonia 72
*Patagotitan* 114, *115*, 155
*Pentaceratops* 50
perching 90

Phytodinosauria 27
phytosaurs 70
piscivory 74, 77, 90, 130
plant-eating *see* herbivory
*Plateosaurus* 30–1, *31*, 58, 153
Portugal 78
predatory dinosaurs 72–3, 78–9, 130–1; *see also* carnivory *and* theropods
predentary 32
prey 134, *135*
*Proceratosaurus* 80, *130*, 130–1, 155
*Protoceratops* 50, 108, 109, 126
*Psittacosaurus* 48, 97, 98, 140
pterosaurs 16, 90, 97
pubis 27, 32, 86

quadrupedalism 30, 32, 41, 43, 50, 54, 57, 60, 124, 128–9, *129*

rauisuchians 22, 35, 70
reproduction 83, 102–5, 106
reproductive organs 110
reproductive rates 119
reptiles, dinosaur descent from 12
research techniques 5, *33*, 78, 98, 136, *136*, 139, *139*
respiratory system 61, 122, 123
rhynchosaurs 22
ribs 43
Romania 68
rostral 48
running 16, 29, 52, 80, 82
Russia 19, 48

saurischians 27
*Saurolophus* 144
sauropodomorphs 24–5, 27, 30–1, *31*, 35, 58, *59*, 60, *117*, 119
sauropods 31, 54, 58, 60–9, 114, *115*
eggs 102, 105
nesting 109
teeth 133
trackways 124
vertebrae 122, *123*
scales 40, 57, 82, 93, *93*
*Scelidosaurus* 41
*Scipionyx* 100, *101*, 155
sclerotic ring 140, *141*
*Scolosaurus* 40, 40–1, *42*, 43, *43*, 153
*Scutellosaurus* 40, 41
Seeley, Harry Govier 26–7
semilunate carpal 86
sexing dinosaurs 110, *112*, 113–14
sexual dimorphism 90
*Shantungosaurus* 56

shoulder joints 86
*Shuvuuia* 140
Siberia 55
*Sinornithosaurus* 94, *94*, 96–7, *97*, 98, 155
*Sinosauropteryx* 80, 80–3, *83*, 94, 97, 98, 113, 134, 154
*Sinosaurus* 71
*Sinraptor* 79
skeletons and replicas 13, *24*, 29, *31*, *37*, 40, *45*, 49, *55*, 56, *64*, 66, *75*, 81, 84–5, 88, 89, 90, *91*, 95, 99, 100, 101, 115, 129
skin 72, 92–3, *93*, 94
skull 23, *23*, 27, *33*, *53*, *57*, *59*, *64*, 78, 79, *119*, *130*
and bite strength *39*, *73*, 80, 131
boxy shape 32, *39*, 66, 72, *73*
crests 33, 57, 70–1, *71*, 82, 142–4, *143*, *145*
domed shape 46–7, *47*
frills 33, 48, 50–1, 112, 128, 145
horns 33, 38, 51, *51*, 72, *73*, 112, 145
spikes *47*, 145
*see also* ear anatomy; eye anatomy; jaw, lower *and* teeth
smallest dinosaur 114
smell, sense of *see* olfactory lobes
social behaviour 142–5; *see also* communal nesting; herding *and* trackways
soft tissues 94, 100, *101*
Sophie (*Stegosaurus*) 44, *45*
sounds, production of 57, 142, 143
South Africa 153, 155
South America 41, 51, 54, 60, 68, 77; *see also* Argentina; Brazil *and* Chile
South Dakota, USA 156
Spain 90
specimens 153–6
composite 113
holotype 53
wartime destruction 75
speed 38, 82, 85, 124, 126
spikes 44, *47*, 55, 145
spinal joints 18
spines 44
*Spinosaurus* 74–5, *75*, 77, *77*, 154
stegosaurs 44–5, 54, 128
*Stegosaurus* 7, 26, 27, 41, 44–5, *45*, 153
Stonesfield slate quarries, UK 11
stride 13, 126

Stromer von Reichenbach, Ernst 74–5
*Struthiomimus* 82
*Styracosaurus* 32, 145
Sue (*Tyrannosaurus*) 110
*Supersaurus* 65
swimming 74–5, *77*, *77*
Switzerland 30, 153
synapsids 22, 35

tail 18, 60
bony struts 80
clubs 41
feathers 88, 90, *91*, 112
and running 52, 80
and sexing dinosaurs 112
spikes 44
and swimming 75, *75*
vertebrae 41, 65, 88
whip-tails 64, *64*, 65
Tanzania 18, 20, 45, 65, 66, 153, 154
teeth 19
canines 38, *39*
carnivores 28, 85, 130, *130*
herbivores 30, 32, 38, 43, 46, 48, 55, 61, 64, 66, 83, 133, *133*
lack of 90
loss 82
numerous 57
piscivores 74, 77, 130
*Teleocrater* 16, *16*, *18*, 18–19, 20, 153
temperature regulation, body 44, 77, 94, 97, 120
tendons, bone 57
*Tenontosaurus* 85
Thailand 48, 77
therizinosaurs 82, 83
theropods 24, 25, 27, 70–1, 74, 78–9, 100, 114, 134, *135*, *141*
body sizes 122
early 28–9
eggs 105
feathered 71, 80–3, 97, 120
nests 108–9
skull *130*, 131
teeth 130, *130*
trackways 124, *125*, 126–7
vertebrae 122
thyreophorans 40–3
*Tianyulong* 97
*Titanoceratops see* Pentaceratops
titanosaurs 68–9, 114, *115*
eggs 102, *102*, *104*, 104–5, 155
trackways 57, 68, 124, *125*, 126–7, 145, 155
*Triceratops* 2, 26, 48, *49*, 50–1, *51*, 114, 134, 145, 150, 153, 154

*Tuojiangosaurus* 128–9, *129*, 155
*Tyrannosaurus* 12, 25, *49*, 51, 77, 80, 82, 110, 114, 118, 128, 130, 131, 134, *135*, 138, *139*, 145, 148–9, *149*, 150–1, *151*, 155

UK 38, 41, *130*
England 11, *15*, *53*, 80, 124, 126, 153, 154, 155, 156
Scotland *127*, 155
Wales *59*, 154
USA 38, 40, 41, 44, 45, 46, 55, 56, 64, 66, 93, 110, *133*
Alaska 120, 147
Arizona 154
Cleveland-Lloyd Dinosaur Quarry 78
Egg Mountain 106, *107*
Ghost Ranch 28, 78
Hell Creek Formation *51*
Montana *51*, 84, 106, *106*, 153, 154
Morrison Formation 7, 78
New Mexico 28, 78, 153
South Dakota 156
Utah 78, 154
Wyoming 84, 106, *106*, 153, 154, 156
Utah 78, 154

*Velociraptor* 25
vertebrae 122, *123*
neck 30, 60–1, 66
processes 74
sacral 14, 20, *21*
tail 41, 65, 88
vertical plates 44–5, *45*
vertical reach 58, 61, 66
vision *see* eye anatomy; eyesight *and* optic lobes
*Vulcanodon* 60

weight *see* body weight
wings 16, 86, 88, 90, *91*
wishbone 28, 88
*Wulong* 98
Wyoming, USA 84, 106, *106*, 153, 154, 156

*Yinlong* 48
young *see* hatchlings *and* juveniles
Yucatan Peninsula, Mexico 149

Zimbabwe 60

# Further reading

Benton, M. J. & Nicholls, R. (2023), *Dinosaur Behavior: An Illustrated Guide*. Princeton University Press.
Brusatte, S. (2019), *The Rise and Fall of the Dinosaurs: The Untold Story of a Lost World*. Picador.
Hone, D. (2022), *The Future of Dinosaurs: What We Don't Know, What We Can and What We'll Never Know*. Hodder & Stoughton.
Naish, D. (2021), *Dinopedia: A Brief Compendium of Dinosaur Lore*. Princeton University Press.
Naish, D. & Barrett, P. (2023), *Dinosaurs: How They Lived and Evolved*, third edition. Natural History Museum, London.
Pickrell, J. (2017), *Weird Dinosaurs: The Strange New Fossils Challenging Everything We Thought We Knew*. Columbia University Press.

# Acknowledgements

Efforts like these are only made possible through the hard work of others. In particular, I'd like to thank David Evans (Royal Ontario Museum) for reading the entire text and picking up on some of my clumsy mistakes, as well as all of those friends and colleagues who donated or suggested images for each of the taxa included, or provided permissions for their use (Kimberley Chapelle, Jonah Choiniere, Ricardo Martínez, Sterling Nesbitt, Emma Nicholls, Rose Prevec, Omar Regalado Fernández, Joep Schaeffer, Daniela Schwarz, Xu Xing). This work has been inspired in part by working with these and other colleagues around the world over the past three decades. Special thanks go to the in-house Publishing team at the Natural History Museum, London, Lesley Simon for her careful copyediting and Lucie Goodayle for some last minute photographs.

# Picture credits

p.7 ©Robert Nicholls / Trustees of the Natural History Museum; p.9, 125 ©Oxford University Museum of Natural History; p.10 ©The Royal Mint / Megalosaurus 2020 UK Coin Collection; p.15, 108, 144 ©John Sibbick/Science Photo Library; p.19 ©Mark Witton/Natural History Museum, London; p.23 ©Museo de Ciencias Naturales, Universidad Nacional de San Juan, Argentina; p.24,25, 26 The Lord of the Allosaurs, CC BY-SA 3.0 via Wikimedia Commons; p.31 ©Joep Schaeffer; p.36, 37 ©Bernhard Zipfel; p.39 ©Prof. Jonah Choiniere/the Albany Museum in Grahamstown, South Africa; p.62/63 ©Júlia d'Oliveira; p.67 ©Shadowgate from Novara, ITALY, CC BY 2.0 via Wikimedia Commons; p.69 ©Mike Taylor; p.71 ©Adam D. Marsh, fossil remains property of the Navajo Nation; p.73 ©Dr Martin Ezcurra/Museo de Ciences Naturales, Buenos Aires; p.74/75 ©3blindMies, CC BY-SA 4.0 via Wikimedia Commons; p.76 ©Mark Garlick/Science Photo Library; p.81, 83, 91 ©The Geological Museum of China/Trustees of the Natural Museum, London; p.85 ©Corbin17/Alamy Stock Photo; p.87 ©Tim Evanson from Washington, D.C., United States of America, CC BY-SA 2.0, via Wikimedia Commons; p.95 ©Martin Shields/Science Photo Library; p.96 ©Emily Willoughby; p.99 (top), 141 ©Prof. Jonah Choiniere/Dr. Xu, Xing/Institute of Vertebrate Paleontology & Paleoanthropology; (bottom) ©Julius Csotonyi/Science Photo Library; p.101 ©Museo Civico di Storia Naturale di Milano; p.111 ©David Varricchio; p.117, 118 ©Kimberley Chapelle; p.121 ©Herne MC, Tait AM, Weisbecker V, Hall M, Nair JP, Cleeland M, Salisbury SW. 2018. A new small-bodied ornithopod (Dinosauria, Ornithischia) from a deep, high-energy Early Cretaceous river of the Australian–Antarctic rift system. PeerJ 5:e4113; p.127 ©Paul M. Barrett; p.134 (top) ©Image courtesy of the Royal Saskatchewan Museum; (bottom) ©François Therrien et al. ,Exceptionally preserved stomach contents of a young tyrannosaurid reveal an ontogenetic dietary shift in an iconic extinct predator.Sci. Adv.9,eadi0505(2023). DOI:10.1126/sciadv.adi0505; p.139 ©WitmerLab at Ohio University; p.148 ©Evolutionnumber9, CC BY-SA 4.0 via Wikimedia Commons.

Unless otherwise stated images copyright of The Trustees of the Natural History Museum, London.

Every effort has been made to contact and accurately credit all copyright holders. If we have been unsuccessful, we apologise and welcome correction for future editions and reprints.